Springer Theses

Recognizing Outstanding Ph.D. Research

For further volumes:
http://www.springer.com/series/8790

Aims and Scope

The series "Springer Theses" brings together a selection of the very best Ph.D. theses from around the world and across the physical sciences. Nominated and endorsed by two recognized specialists, each published volume has been selected for its scientific excellence and the high impact of its contents for the pertinent field of research. For greater accessibility to non-specialists, the published versions include an extended introduction, as well as a foreword by the student's supervisor explaining the special relevance of the work for the field. As a whole, the series will provide a valuable resource both for newcomers to the research fields described, and for other scientists seeking detailed background information on special questions. Finally, it provides an accredited documentation of the valuable contributions made by today's younger generation of scientists.

Theses are accepted into the series by invited nomination only and must fulfill all of the following criteria

- They must be written in good English.
- The topic of should fall within the confines of Chemistry, Physics and related interdisciplinary fields such as Materials, Nanoscience, Chemical Engineering, Complex Systems and Biophysics.
- The work reported in the thesis must represent a significant scientific advance.
- If the thesis includes previously published material, permission to reproduce this must be gained from the respective copyright holder.
- They must have been examined and passed during the 12 months prior to nomination.
- Each thesis should include a foreword by the supervisor outlining the significance of its content.
- The theses should have a clearly defined structure including an introduction accessible to scientists not expert in that particular field.

Karl Schmid

Laser Wakefield Electron Acceleration

A Novel Approach Employing Supersonic Microjets and Few-Cycle Laser Pulses

Doctoral Thesis accepted by
Max Planck Institute for Quantum Optics
Garching, Germany

Springer

Author
Dr. Karl Schmid
Max Planck Institute for Quantum Optics
Hans-Kopfermann-Str. 1
85748 Garching
Germany
e-mail: karl.schmid@gmx.eu

Supervisor
Prof. Ferenc Krausz
Max Planck Institute for Quantum Optics
Hans-Kopfermann-Str. 1
85748 Garching
Germany
e-mail: ferenc.krausz@mpg.de

ISSN 2190-5053 e-ISSN 2190-5061

ISBN 978-3-642-19949-3 e-ISBN 978-3-642-19950-9

DOI 10.1007/978-3-642-19950-9

Springer Heidelberg Dordrecht London New York

Cover design: eStudio Calamar, Berlin/Figueres

Printed on acid-free paper

Springer is part of Springer Science+Business Media (www.springer.com)

Publications by the Author

F. Tavella, K. Schmid, N. Ishii, A. Marcinkevičius, L. Veisz, and F. Krausz. Highdynamic range pulse-contrast measurements of a broadband optical parametric chirpedpulse amplifier. Appl. Phys. B, 81:753, 2005.

T. Metzger, J. Rauschenberger, A. Apolonski, K. Schmid, A. Baltuška, and F. Krausz. Ultrakurze laserpulse. Laser Technik J., 4:27, 2005.

A. J. Verhoef, J. Seres, K. Schmid, Y. Nomura, G. Tempea, L. Veisz, and F. Krausz. Compression of the pulses of a ti:sapphire laser system to 5 femtoseconds at 0.2 terawatt level. Appl. Phys. B, 82:513–517, 2006.

A. Stalmashonak, N. Zhavoronkov, I. V. Hertel, S.Vetrov, and K. Schmid. Spatial control of femtosecond laser system output with submicroradian accuracy. Appl. Opt., 45(6):1271–1274, 2006.

B. Hidding, K.-U. Amthor, B. Liesfeld, H. Schwoerer, S. Karsch, M. Geissler, L. Veisz, K. Schmid, J. G. Gallacher, S. P. Jamison, D. Jaroszynski, G. Pretzler, and R. Sauerbrey. Generation of quasimonoenergetic electron bunches with 80-fs laser pulses. Phys. Rev. Lett., 96(10):105004, 2006.

N. Ishii, C. Y. Teisset, T. Fuji, S. Köhler, K. Schmid, L. Veisz, A. Baltuška, and F. Krausz. Seeding of an eleven femtosecond optical parametric chirped pulse amplifier and its Nd3+ picosecond pump laser from a single broadband ti:sapphire oscillator. IEEE J. Quant. Elec., 12(2):173, 2006.

Y. Nomura, L. Veisz, K. Schmid, T. Wittmann, J. Wild, and F. Krausz. Time-resolved reflectivity measurements on a plasma mirror with few-cycle laser pulses. New J. Phys., 9(1):9, 2007.

B. Hidding, G. Pretzler, M. Clever, F. Brandl, F. Zamponi, A. Lübcke, T. Kämpfer, I. Uschmann, E. Förster, U. Schramm, R. Sauerbrey, E. Kroupp, L. Veisz, K. Schmid, S. Benavides, and S. Karsch. Novel method for characterizing relativistic electron beams in a harsh laser-plasma environment. Rev. Sci. Inst., 78(8):083301, 2007.

S. Karsch, J. Osterhoff, A. Popp, T. P. Rowlands-Rees, Zs. Major, M. Fuchs, B. Marx, R. Horlein, K. Schmid, L. Veisz, S. Becker, U. Schramm, B. Hidding, G. Pretzler, D. Habs, F. Gruner, F. Krausz, and S. M. Hooker. Gev-scale electron acceleration in a gas-filled capillary discharge waveguide. New J. Phys., 9(11):415, 2007.

K. Schmid, L. Veisz, F. Tavella, S. Benavides, R. Tautz, D. Herrmann, A. Buck, B. Hidding, A. Marcinkevicius, U. Schramm, M. Geissler, J. Meyer ter Vehn, D. Habs, and F. Krausz. Few-cycle laser-driven electron acceleration. Phys. Rev. Lett., 102(12):124801, 2009.

L. Veisz, K. Schmid, F. Tavella, S. Benavides, R. Tautz, D. Herrmann, A. Buck, B. Hidding, A. Marcinkevicius, U. Schramm, M. Geissler, J. Meyer ter Vehn, D. Habs, and F. Krausz. Laser-driven electron acceleration in plasmas with few-cycle pulses. Compt. Rend. Phys., 10(2-3):140–147, 2009.

B. Hidding, M. Geissler, G. Pretzler, K.-U. Amthor, H. Schwoerer, S. Karsch, L. Veisz, K. Schmid, and R. Sauerbrey. Quasimonoenergetic electron acceleration in the selfmodulated laser wakefield regime. Phys. Plasmas, 16(4):043105, 2009.

A. Buck, K. Zeil, A. Popp, K. Schmid, A. Jochmann, S. D. Kraft, B. Hidding, T. Kudyakov, C. M. S. Sears, L. Veisz, S.Karsch, J. Pawelke, R. Sauerbrey, T. Cowan, F. Krausz, and U. Schramm. Absolute charge calibration of scintillating screens for relativistic electron detection. Rev. Sci. Inst., 81:033301, 2010.

C. M. S. Sears, S. Benavides Cuevas, U. Schramm, K. Schmid, A. Buck, D. Habs, F. Krausz, and L. Veisz. A high resolution, broad energy acceptance spectrometer for laser wakefield acceleration experiments. Rev. Sci. Inst., 81:073304, 2010.

C. M. S. Sears, A. Buck, K. Schmid, J.M. Mikhailova, F. Krausz, and L. Veisz. Emittance and divergence of laser wakefield accelerated electrons. Phys. Rev. Spec. Top. Acc. Beam., 13:092803, 2010.

K. Schmid, A. Buck, C. M. S. Sears, J.M. Mikhailova, R. Tautz, D. Herrmann, M. Geissler, F. Krausz, and L. Veisz. Density-transition based electron injector for laser driven wakefield accelerators. Phys. Rev. Spec. Top. Acc. Beam., 13:091301, 2010.

Supervisor's Foreword

Particle accelerators play an indispensable role in several branches of modern research such as, inter alia, physical, biological and materials research or medicine. In many of these applications and using current particle accelerator technology, these devices are very large and costly thereby impeding installation at small and medium-scale research institutes or universities. The alternative approach of laser-driven electron acceleration, which is the object of research of the present thesis, holds the promise of reducing the size and cost of conventional accelerators by many orders of magnitude thereby making this indispensable scientific tool available to a much broader range of users. Additionally, laser driven electron acceleration offers the advantage of generating electron pulses which are significantly shorter than those produced by conventional technology. This unique property has the potential of enabling ground-breaking research of certain ultra-fast processes with unprecedented temporal resolution.

In the present thesis two separate topics are investigated, with both relating to electron acceleration facilitated by the strongly relativistic interaction of an ultra-intense laser pulse with a fully ionized helium plasma. Firstly, microscopic supersonic gas jets necessary for the experimental realization of the acceleration process are explored. Detailed theoretical analysis, design and characterization of nozzles producing microscopic supersonic gas jets are presented in part one of this thesis. These investigations lead to the generation of the smallest tailored gas targets ever used in the context of laser-plasma research.

Secondly, the generation of ultra-short, relativistic electron bunches is investigated. The generation, precise optimization and characterization of the accelerated electron bunches are the main topics of part two of this thesis. Stable monoenergetic electron beams were produced in a complex experimental setup including a high degree of freedom in the relevant parameter ranges and with a unique laser system producing ultra-short and ultra intense pulses.

In order to characterize the properties of the electron bunches, several electron beam diagnostics were designed and used for detailed studies of various properties of the electron bunches such as energy spectrum, angular distribution, divergence and charge. The acceleration experiments are performed with high repetition rate and deliver a pulsed electron beam with an electron energy spectrum free of low-energetic background. Careful parameter scans using a large number of samples yielded the optimal experimental parameters and valuable information about the physics of the laser-plasma interaction.

The knowledge gained from this work is a solid basis of forthcoming research. One of these directions is the development of a novel injector placing electrons into the accelerating plasma wave thereby further stabilizing the acceleration process. Another outcome is new characterization techniques of unexplored electron beam properties such as normalized transverse emittance, electron bunch duration or the structure of the electron plasma wave accelerating the particles. Therefore, this work is not only a detailed universal description of supersonic nozzles for a broad field of applications but also a pioneering investigation on plasma based laser-driven electron acceleration.

Garching, January 2011 Ferenc Krausz

Acknowledgments

Laser-plasma physics is one of the areas of experimental research where only a team of researchers is able to get the experiment to work and, finally, the paper published. This was certainly the case in the present work and I am deeply indebted and grateful to the following persons:

First, I would like to thank my doctoral advisor Prof. Dr. Ferenc Krausz for giving me the opportunity to work as a doctoral student in his group. I would also like to thank him for his advice and encouragement that were always at hand when needed.

Special thanks go to Dr. Laszlo Veisz for all the effort he put into supervising my work on this thesis. He certainly was not only a great supervisor but also a dependable colleague in the laboratory where he stayed many long nights with us.

Prof. Dr. Toshiki Tajima was kindly willing to review my thesis. I am very grateful for and honored by the fact that the very inventor of laser-driven electron acceleration is in this way also part of this work.

I would like to thank Prof. Dr. Jürgen Meyer-ter-Vehn for his enthusiasm in laser plasma physics that also sparked my interest. I also would like to thank him for many interesting discussions.

In the lonely field of gas dynamics at the MPQ, Dr. Hartmut Schröder was one of the very few persons with experience in this area. I thank him for sharing his experience with me.

I thank Dr. Andrius Marcinkevičius for all the work he put into the development the laser system employed in the experiments.

Special thanks goes to my colleague as a doctoral student Dr. Franz Tavella. Not only was he tirelessly working for getting the laser ready for the experiment. He also spent many a night with me in the lab, accelerating electrons.

All the PIC simulations that helped us so much in understanding the physics of our electron accelerator were done by Dr. Michael Geissler with his PIC-code ILLUMINATION. I am very grateful for all the time and energy he invested in producing these results.

Many thanks goes to Sofia Benavides who put so much work into the fibre-based electron detection of the electron spectrometer.

To Daniel Hermann I would like to extend my thanks for his efforts of constantly making the pulses shorter and putting more energy into them.

I thank Raphael Tautz for his contributions to the experiment as a diploma student as well as for his efforts in laser development as a doctoral student.

I would like to thank Alexander Buck and Dr. Christopher Sears for being such reliable partners in the electron acceleration business. I am also grateful that they gave me the time I needed for writing up my thesis.

Manfred Fischer, Alois Böswald, Harald Haas, and Hans-Peter Schönauer were involved in the design and construction of many parts and devices of the experiment. I am very grateful for their contributions and for their help which they always offered when it was needed.

I would also like to express my thanks to the workshop staff who never turned me down although I had many special and time consuming requests.

Of course, without the support and trust of my family, I would not have been able to accomplish this work. Therefore, special thanks goes to the following people:

First, I would like to express my gratitude to TR Dr. Hans Kaltenbrunner for so many years of support.

My fiancé, Andrea Wechsler, I would like to thank for her consideration as well as her active support during the time I was writing this thesis. I also thank her for proofreading the manuscript.

Finally I would like to offer my special thanks to my parents, for their support and encouragement throughout my study.

Contents

Chapter 1
Introduction

1.1 Motivation

For a century, the on-going development of particle accelerators has been promoting many branches of fundamental and applied research. What began as a tool for nuclear and particle physics, has expanded its use into solid state physics as well as medicine, biology and even history [1]. As these lines are written, the superconducting magnets of the Large Hadron Collider (LHC) [2–8] at the CERN laboratory are being cooled down to liquid Helium temperature and in a few months' time, the largest collider ever built will commence operation. With its two counter-propagating proton beams having 7 TeV energy each, it is expected to shed new light on hot topics such as the fundamental origin of mass in form of the famous HIGGS Boson [9], dark energy and dark matter [10], the possible existence of small extra dimensions in space-time [11], and many more. However, looking at the tremendous scale of this project, it is valid to ask the question whether this collider will actually *stay* the largest collider ever built for many decades to come. With the Superconducting Super Collider (SSC) [12–15] in Texas, USA, having been cancelled in 1993 due to exploding cost-forecasts that saw the final price tag exceeding 12 billion USD, the only remaining accelerator project which is of comparable magnitude to the LHC is the International Linear Collider (ILC) [16, 17]. The latter will—if realized—consist of two linear accelerators, in head-on configuration, one accelerating electrons, the other one positrons. The entire structure will stretch over a length of 31 km and will be able to reach a particle energy of 500 GeV in each beam. With a projected total cost of 5 billion USD, it can only be realized by an international collaboration of several contributing countries.

One of the reasons for these impressive spatial and financial dimensions is a basic physical limitation of conventional accelerator technology: breakdown [18] occurring at the walls of the accelerating structure at field strengths of roughly 100 MV/m. For example, the accelerating structures planned to be used in the ILC will support an accelerating gradient of approximately 32 MV/m. Therefore, in order to reach the TeV-scale in linear configuration, which is imperative in the

K. Schmid, *Laser Wakefield Electron Acceleration*, Springer Theses,
DOI: 10.1007/978-3-642-19950-9_1, © Springer-Verlag Berlin Heidelberg 2011

case of electrons and positrons due to prohibitive radiation losses in synchrotron configuration [19], the accelerator needs to be more than 30 km long.

This is where plasma based electron accelerators offer great advantages due to their orders of magnitude larger accelerating gradients which in turn allow for much shorter acceleration length. In this accelerator concept a laser- or ultra-relativistic electron pulse is used to excite a plasma wave which then serves as accelerating structure. As the pulse propagates through the plasma, it excites a running, longitudinal charge separation wave—called Langmuir wave—that trails the electron or laser pulse and, therefore, has a phase velocity equal to the group velocity of the laser pulse or equal to the velocity of the electron pulse. Depending on plasma density and electron temperature, this type of wave generates longitudinal electric fields in the 1 GV/m to 1 TV/m range. This concept of using a travelling plasma wave as an accelerating structure was first presented by Tajima and Dawson in 1979 when they proposed to excite such a plasma wave by means of a short and powerful laser pulse [20].

1.2 Plasma-Based Electron Acceleration: A Historical Overview

With the prospects of this new concept as a motivation, diverse schemes to excite longitudinal plasma waves and to use them for electron acceleration were experimentally investigated, the most important ones being the plasma wakefield accelerator (PWFA) [21–26], the plasma beatwave accelerator (PBWA) [27–31], and the laser wakefield accelerator LWFA [32–47].

In the case of PWFA, the plasma wave is excited by one or more relativistic electron bunches and another pre-accelerated bunch is then injected and accelerated by the plasma wave. For many years, maximum energy gains were quite moderate in this field [21–24] with a maximum gain of 30 MeV of an electron bunch with an initial energy of 500 MeV [23]. A more complete account of the developments in this field up to the year 1995 can be found in [48]. When improving electron bunch parameters allowed better matching of electron bunch size and length to the plasma density, substantial energy gain in the GeV range was demonstrated [25] and the so-called plasma-afterburner was recently shown to be capable of energy-doubling the trailing edge of a short and dense 42 GeV electron bunch in a meter-scale plasma [26].

The concept of PBWA was proposed by Tajima and Dawson [20] as a work-around solution to drive a plasma wake-field without the necessity of ultrashort, high-power lasers, which only became available after the invention of the chirped-pulse amplification technique in 1985 [49]. Two long-pulse (ns) laser beams with slightly different wavelength are spatially overlapped in a plasma. The wave-length-difference is tuned such that the running beat-wave resonantly excites a plasma wave which is then used to accelerate injected electrons [27–31]. Energy gains up to 28 MeV have been shown in this scheme [28, 29].

As powerful ultrashort pulse lasers became available, the initial concept of laser wakefield acceleration could be realized. In LWFA, a short ($\tau < 1\mathrm{ps}$), ultra-intense ($I > 10^{18}$ W/cm^2) laser pulse drives the plasma wave. The plasma wave excitation is most efficient, when the laser pulse length is approximately equal to or shorter than the plasma wavelength. The first experimental evidence of the excitation of a plasma wave in the LWFA regime was found in 1993 by Hamster et al. [32] and two years later also by other groups [34, 35]. In 1995, electron acceleration by LWFA was shown by injecting electrons emitted from a laser-solid interaction into a low-density plasma wave by Nakajima et al. [33]. In this case, an 8 TW, 1 ps laser pulse impinged on a plasma with a density of 3.5×10^{15} cm^{-3} thereby exiting a plasma wave that accelerated the injected electrons from an initial energy of 1 MeV to a final one of 8 MeV corresponding to an estimated accelerating gradient of 0.7 GV/m.

As mentioned previously, in principle much larger accelerating fields are feasible but they require the plasma wave to be driven much stronger than it was possible in the early days of CPA by a single pulse that was resonant with the plasma wave. Therefore, the self-modulated laser wakefield (SM-LWFA) scheme was proposed [50–52]. As the standard LWFA, the driving laser pulse has a sub-ps duration and an intensity typically larger than 10^{18} W/cm^2 but the plasma density is larger here such that the laser pulse length is (much) longer than the plasma wavelength. The plasma density is typically chosen such that the laser pulse power is a little bit larger than the critical power for relativistic self-focusing [53–57]. In this case, self guiding occurs which compensates diffraction and, therefore, causes the laser pulse to maintain or surpass its focal intensity for a distance that can greatly exceed the Rayleigh-range of the beam. The Rayleigh-range is the on-axis distance from the laser focus where the laser intensity is reduced to half its value in the focal spot due to diffraction. This prolonged interaction with the plasma at ultrahigh intensity allows for the growth of the self-modulation or beam-breakup instability which splits the laser pulse up into multiple beamlets each of which then drives the plasma wave resonantly. In this way, highly anharmonic Langmuir waves can be excited leading to enhanced acceleration gradients. The beam-breakup can be described in terms of the forward Raman-scattering instability [58, 59] where the strong electromagnetic pump wave (ω, k) decays into a plasma wave (ω_p, k_p) and two forward-propagating electromagnetic waves at the so-called Stokes $(\omega - n\omega_p)$ and anti-Stokes $(\omega + n\omega_p)$ frequencies. Here n is an integer enumerating the sidebands. The detection of the Stokes and anti-Stokes sidebands serves in many experiments as a proof of the SM-LFWA process and can also be used for measuring the plasma density in the interaction region due to the side-band-spacing dependance on the plasma frequency.

First experimental evidence for wake field excitation in this regime was presented by Coverdale et al. in 1995 [40]. Again, Nakajima et al. were the first to show acceleration in this regime in the same year [41] by injecting and accelerating a 1 MeV electron bunch up to 17 MeV with an estimated acceleration gradient of 30 GV/m. They used a 3 TW, 1 ps, 10^{17} W/cm^2 pulse in a plasma with

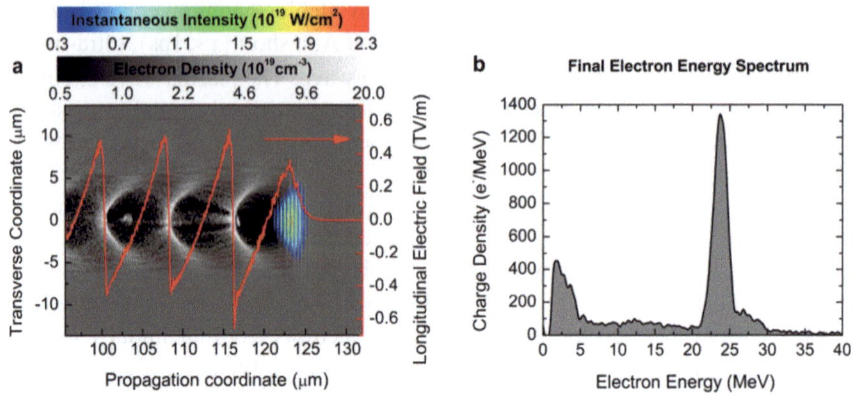

Fig. 1.1 Particle-In-Cell (PIC) simulation of electron acceleration in the bubble regime with 40 mJ, 8.5 fs (FWHM) laser pulses. **a** shows the physical state (plasma density, *grey-scale* contour, instantaneous laser intensity, rainbow contours and longitudinal electric field, *red line*) of the accelerator after 123 μm propagation. **b** is the electron energy spectrum after the accelerated electron bunch propagated out of the plasma. Simulations courtesy of Michael Geissler et al. [61]

an electron density of $10^{19}\,\mathrm{cm}^{-3}$. Subsequently, electron acceleration in the SM-LWFA regime was realized in many experiments [42–46]. Modena et al. were the first to show substantial wave breaking in the self modulated regime which was accompanied by clearly non-maxwellian spectra with maximum energies around 44 MeV. Gahn et al. [60] showed that in the laser and plasma parameter region of SM-LWFA, a competing electron acceleration process termed direct laser acceleration (DLA) exists which under certain circumstances dominates over wakefield acceleration. The largest electron energy in SM-LWFA as well as in the DLA regime so far was shown in 2005 by Mangles et al. using a petawatt laser where exponentially decaying energy spectra and spectra following a power law with a maximum energy larger than 300 MeV were shown. Common to all the above mentioned experiments is the large electron energy spread that makes the generated electron beams of limited use for application.

Meanwhile, with continuing laser development, Ti:sapphire based CPA lasers reached ever shorter pulse durations with increasing pulse energies. Eventually this allowed experimenters to go back to the initial LWFA concept as the short pulse durations now allowed the use of a pulse short enough to fit into one plasma wavelength and still drive the wave strong enough to reach wave breaking. On the border between LWFA and SM-LWFA, some authors, such as Malka et al. in 2002, speak of the "forced-LWFA"-regime [47] where these new ultrashort $\sim$40 fs and relativistically intense $>10^8$ W/cm^2 pulses are used in plasmas with a density around $10^{19}\,\mathrm{cm}^{-3}$ so that still several plasma wavelength fit into the laser pulse. While this regime produced pedestal-type spectra that were clearly non-maxwellian [47] it failed to produce monoenergetic spectra that are demanded by most applications (Fig. 1.1).

Such monoenergetic spectra were predicted by Pukhov and Meyer-ter-Vehn in 2002 in their work about three dimensional plasma waves which are so strongly driven that they transversally break. In this regime, the laser intensity is so large that the generated plasma wave breaks directly behind the pulse and some electrons of the background plasma are injected into the first wake of the plasma wave and are accelerated. Numerical studies [62–64] have shown that the accelerated electrons emerge from the plasma as monoenergetic electron bunches with relativistic energy and few-femtosecond duration. Under optimal conditions, the driving laser pulse has relativistic intensity $>10^8$ W/cm^2 and a duration and diameter that are matched to the plasma density. This demand calls for a pulse-length equal to or less than half the plasma wavelength, and a focal diameter approximately equal to the plasma wavelength. In this case, the ponderomotive force of the laser pulse is so large that the majority of the free electrons are transversally pushed out leaving the positively charged ions behind. After a propagation length comparable to a plasma wavelength, the electrons are driven back to the axis by the fields built up by charge separation. A cavity void of electrons trailing the laser pulse emerges: it has been dubbed a “bubble” [62]. A fraction of the returning electrons is injected and trapped in the bubble and accelerated by its strong longitudinal electric field resulting in relativistic electron bunches with narrow-band energy spectra [65, 66].

Only recently laser pulses with parameters coming close to the requirements of the bubble regime—like in the present work and in Hafz et al. [67]—became available, so it came as a surprise when in 2004 three independent groups showed the generation of monoenergetic electron beams, namely Faure et al. [68], Geddes et al. [69], and Mangles et al. [70]. In all the three cases the employed laser pulses did not initially match the requirements prescribed by the so-called "Bubble-Regime" by Pukhov and Meyer-ter-Vehn. PIC simulations conducted with the exact experimental parameters [68–70] showed that self-focusing and self-compression of the laser pulses in the plasma finally transformed the pulses into the required domain.

In the last few years, a number of studies have been devoted to laser-generated monoenergetic electron beams and all of them—with the exception of the results presented by Hidding et al. [71]—work in the above described regime where the laser pulse initially (almost) fits into one plasma wavelength but only through some self-modulation acquires high enough intensity to drive the wake wave into breaking. Leemans et al. demonstrated the record electron energy of 1 GeV [72], using a discharge in a gas-filled capillary to guide the 40 fs, 40 TW laser pulse with an initial intensity of 3×10^{19} W/cm^2 over a cm-scale distance in a 3×10^{18} cm^{-3} plasma and, thereby, maximizing the acceleration length. Faure et al. [73] showed that by using a second weaker, counter-propagating laser pulse that was overlapped with the primary driver pulse in a 7.5×10^{18} cm^{-3} plasma injection of background electrons into the plasma wave could be triggered without actually driving the plasma wave to breaking. This leads to a greatly improved shot-to-shot reproducibility of the monoenergetic electron beam and also allows

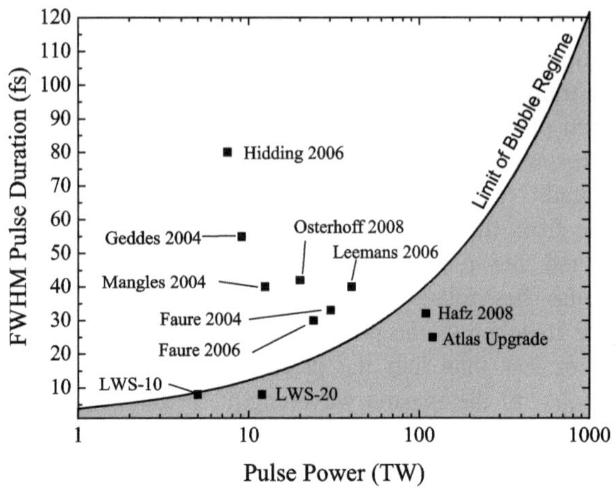

Fig. 1.2 Pulse duration limit (fs) of the bubble regime versus pulse power (TW) (*solid line*). The allowed parameter range for the bubble regime is shaded in grey. Parameters of former and present experiments (*black rectangles*). The following experiments are included: Mangles 2004 [70], Geddes 2004 [69], Malka 2004 [68], Leemans 2006 [72], Faure 2006 [73], Hidding 2006 [71], Osterhoff 2008 [75], Hafz 2008 [67]. LWS-10 corresponds to the present work and LWS-20 to the parameter of an upgrade of the system that is under way. "Atlas Upgrade" corresponds to the planned parameters of the upgraded Ti:sapphire laser system ATLAS at the Max-Planck-Institute of Quantum Optics

tuning of the injected charge and electron energy [74]. Both laser pulses had 30 fs duration, the driving laser with 24 TW power and the injection laser with 8.3 TW. Osterhoff et al. [75] showed improved shot-to-shot reproducibility by employing a steady-state gas-cell as a target and Hafz et al. [67] also showed improved stability by using ultra-high power (110 TW) ultrashort (32 fs) laser pulses. As a summary to the LWFA experiments that generated monoenergetic electron pulses, Fig. 1.2 shows the relation of pulse duration versus pulse power of the above cited experiments together with the threshold that defines the bubble regime. Here, the grey area is within and the white area is outside this regime. The figure clearly shows that the present work is together with Hafz et al. the only one that falls into this regime and that it is far removed from all other experiments thereby exploring an entirely new parameter range.

1.3 The Present Electron Acceleration Work

In the present work, monoenergetic electron acceleration is demonstrated in a new laser-parameter range by employing for the first time few-cycle laser pulses to LWFA in the bubble regime. These pulses were produced by a novel laser system, LWS-10 (Light Wave Synthesizer-10), the world's first multi-TW sub-10-fs light source [76]. It draws on a non-collinear optical parametric chirped pulse amplifier allowing the amplification of broad bandwidth pulses. In our investigations, the system produced pulses with 50 mJ energy, 8 fs duration and spectra covering the range of 700–980 nm at a 10 Hz repetition rate. This allowed direct access to the bubble-regime. Analytical scaling laws [63–65] and our simulations [61] indicate that for laser pulses shorter than 10 fs, the bubble regime can be accessed with less than 100 mJ pulse energy. Thanks to a laser pulse duration of 8 fs, an on-target laser pulse energy as low as 40 mJ enabled the acceleration process to work giving rise to clean, monoenergetic electron spectra in the range of several 10 MeV and with a divergence of approximately 5 mrad.

1.4 Motivation for the Development of Microscopic de Laval Nozzles

The unique laser properties at hand in the present work also necessitate unique plasma parameters. Our simulations [61] show that very large acceleration gradients in the 1 TV/m range are generated but due to the small laser pulse energy of 40 mJ , the laser pulse can only sustain a stable plasma-bubble over propagation distances in the 100 μm range. As soon as the bubble-structure becomes instable, the electron beam quality deteriorates so that any further propagation inside the plasma must be avoided. Plasma lengths in the 100 μm range have to be used in order to obtain a clean and stable electron beam. The development of nozzles producing supersonic gas jets in this size range was, therefore, an important part of this work. Supersonic flows in microscopic nozzles show a strongly modified behavior as compared to macroscopic flows. This is mainly due to the fact that viscosity and turbulence effects are negligible for most supersonic flows down to the mm size range. In microscopic flows, however, these effects can not be neglected any more and in some cases are even dominant. Since no studies of nozzle design in the context of laser-plasma physics are available in this microscopic regime, a detailed study on this topic based on numeric flow simulation was conducted within the present work.

1.5 Supersonic (Micro) Gas Jets

A central problem in laser-plasma physics is posed by the fact that high power lasers, with laser pulse powers typically larger than 1 TW, cannot be focused in air or any other ambient medium without being strongly modified by ionization of and subsequent nonlinear interactions with the medium leading to strong spatial and temporal pulse modulation and beam breakup. Therefore, in order to reach the intensities of more than 10^{18} W/cm^2 typical in such experiments, the laser must be brought to its focus in an evacuated chamber. In order to obtain clean interaction conditions, a gas target is needed that is spatially well confined so that interaction happens only at the ultrahigh intensity laser focal spot. Several different methods are used in this context in order to obtain gas densities that typically lie between 10^{18} and 10^{20} 1/cm^3. The easiest way would certainly be a simple gas filled chamber with windows for the laser but of course this cannot be used because the windows would immediately be destroyed by the laser. Therefore, either gas jets that emanate into the vacuum or gas cells with holes allowing the laser to pass through or gas filled capillaries with open ends have to be used. All of these methods have in common that a substantial gas load is produced which has to be supported by the vacuum pump system. Many laser-plasma interactions are strongly nonlinear processes and plasma parametric instabilities make the interaction dynamics depend strongly on the exact density profile of any given target and on its shot-to-shot reproducibility. This makes tailored and well characterized gas targets an essential prerequisite for these experiments.

Supersonic gas jets represent a unique tool to produce reproducible, spatially well confined gas targets which offer the possibility of producing flat-top density profiles and steep gradients at the jet edges. Therefore, they are employed in a great number of laser-plasma experiments. The types of experiments are manyfold, a recently very popular one is laser driven electron acceleration [48, 67–71, 73, 77–85]. Others include generation of THz [86–88], X-UV [89] and X-ray [90–94] beams as well as ion beams [95, 96], and laser pulse compression [97] or laser frequency upshift [98] and the generation of strong, quasistatic magnetic fields [99, 100].

Much theoretical and experimental work has been done concerning nozzles generating cold gas jets as a source of clusters [101–109]. Since clustering takes some time to occur, however, the region of interest in these studies is usually far away from the nozzle orifice—many times the orifice diameter. In contrast, due to the specific requirement of rather high particle densities in the context of laser plasma experiments, the region of interest of a gas jet emanating into a vacuum always lies very close to the exit of the gas nozzle. Several previous studies exist about the production and characterization of nozzles producing gas targets in the above mentioned density range—and beyond it. They include different measurement techniques for the characterization of the gas jet like electron diffraction [110], nuclear scattering [111], stimulated Raman scattering [112], white light generation [113], linear Raman spectroscopy [114], and, most frequently used for high density jets, interferometrically [115–120]. These studies present measurements for a small number of nozzle designs used in the experiments. Some authors published work on the optimization of a specific nozzle design via numeric flow simulation [121, 122]. To the authors' knowledge, only one work exists representing first steps towards the systematic numeric and experimental optimization of the properties of mm-size supersonic gas jets close to the nozzle exit for laser-plasma experiments [121]. No study exist so far on the specific behavior of supersonic flows inside microscopic nozzles in the context of laser-plasma physics.

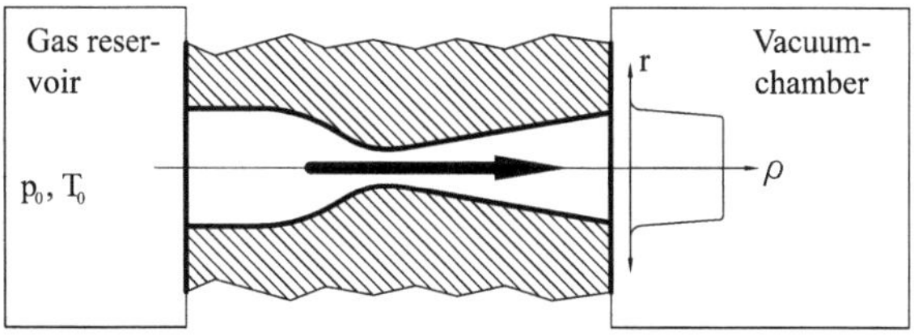

Fig. 1.3 Sketch of a converging-diverging de Laval nozzle as it was used in the present simulations and experiments. On the *left side*, the nozzle is attached to a gas reservoir with a pressure p_0 of typically between 10 and 100 bar and with a temperature T_0 of 300 K. The gas is allowed to expand through the nozzle into an evacuated chamber. In the diverging section of the nozzle after the nozzle throat supersonic conditions are reached. The *arrow* indicates the direction of the gas flow. The supersonic jet emanating into the vacuum has low divergence, a flat top density profile and sharp edges. This is shown in the sketched density profile showing a line-out of the gas density ρ perpendicular to the jet axis versus the jet radius r

Part I of this work reports on the systematic simulation, design, and characterization of supersonic gas jet targets for electron acceleration with ultra-high intensity ultra-short pulse lasers. As detailed above, the laser-parameters at hand dictate the use of very small gas jet diameters in the 100 μm range. At the same time, the fact that the laser focal spot can, depending on focusing conditions, be in the range of tens of micrometers causes the additional requirement that the gas jet should stay collimated and sharp-edged and maintain its flat top profile over a distance of up to one nozzle exit diameter. It will be shown that supersonic gas jets from correctly designed de Laval nozzles fulfill all these requirements in a wide parameter range in terms of gas density and nozzle size. A sketch of such a nozzle is depicted in Fig. 1.3.

The feasibility of microscopic supersonic jets has so far mainly been investigated in the context of micro-thrusters for small scale spacecraft, simulations being done mostly in 2D owing to the geometry of those thrusters [123, 124]. In this context, many publications exist on comparison and benchmarking of different flow models for rarefied flows [125–132], on the optimization of thrusters for microelectromechanical systems (MEMS) [133–136], on properties of jets emanating into surrounding atmosphere [137] and on the fabrication of de Laval micronozzles [138]. The main goal in these publications is to maximize the efficiency and hence the thrust per mass unit of propellant and not the flow profile after the nozzle exit.

In order to obtain optimal target conditions in the present experiments, a systematic numerical study of supersonic gas jets for laser-plasma experiments was conducted. De Laval nozzles with 3 mm to 150 μm exit diameter were then designed and constructed. As far as the smallest nozzles are concerned, the latter point alone proved to be a formidable challenge for state-of-the-art manufacturing technologies. The gas jets were then characterized and finally successfully employed in laser-plasma experiments.

1.6 Structure of the Thesis

The present work is structured in two parts:

Part I, containing Chaps. 2 through 4, is devoted to an in-depth study of supersonic gas flows in microscopic nozzles:

Chapter 2 summarizes the most important and relevant parts of the theory of compressible fluid flow. It starts with basic concepts, such as the equation of state and different types of state changes, then introduces several effects like shock front formation and propagation of distortions at supersonic speed which are peculiar in supersonic gas flows. Cluster formation in supersonic gas flows is also covered in so far as it is relevant to the present work.

Chapter 3 introduces numeric flow simulation in the continuum regime as well as some concepts like boundary layer formation and flow rarefication that are important for microscopic flows. It then summarizes the results of simulations that were conducted in order to understand and optimize the functionality of microscopic gas nozzles for laser-plasma experiments.

Chapter 4 presents experimental results about the characterization of microscopic supersonic gas jets. It is shown that simulation and measurement agree well and that supersonic gas jets with diameters as small as 150 μm can be produced by microscopic de Laval nozzles. It also shows the first experimental steps towards gas targets with structured density profiles.

Part II, spanning Chaps. 5 through 8, covers electron acceleration in relativistic plasma waves driven by few-cycle laser pulses:

Chapter 5 summarizes the analytical and simulation results that are relevant for laser-driven electron acceleration. This includes important effects in analytical description as wake field formation, relativistic self-focusing of laser-pulses, and wave breaking. Empirical laws derived from particle-in-cell (PIC) simulations are discussed as well as the results of PIC simulations matching the experimental conditions in the present laser driven accelerator.

Chapter 6 contains a description of the experimental setup, covering the light-source and the actual electron accelerator.

Chapter 7 gives an account of the main experimental results obtained in the course of this work. The typical electron beam parameters are summarized and results concerning stability and tunability of the acceleration process are presented. A detailed discussion of the experimental results is given.

Chapter 8 discusses the most important problems that need to be solved in the future and presents several promising routes towards the solution of these problems.

Chapter 9 concludes the thesis by summarizing the most important results.

References

1. Dik, J., Janssens, K., VanDer Snickt, G., van der Loeff, L., Rickers, K., Cotte, M.: Visualization of a lost painting by vincent van gogh using synchrotron radiation based X-ray fluorescence elemental mapping. Anal. Chem. **80**(16), 6436–6442 (2008)
2. Evans, L., Bryant, P.: LHC machine. J. Instrum. **3**(08), S08007 (2008)
3. Anelli, G., et al.: The TOTEM Collaboration, The totem experiment at the CERN large hadron collider. J. Instrum. **3**(08), S08007 (2008)
4. Adriani, O., et al.: The LHCf Collaboration, The LHCf detector at the CERN large hadron collider. J. Instrum. **3**(08):S08006 (2008)
5. Alves, A. A. Jr., et al.: The LHCb Collaboration, The LHCb detector at the LHC. J. Instrum. **3**(08):S08005 (2008)
6. Chatrchyan, S., et al.: The CMS Collaboration, The CMS experiment at the CERN LHC. J. Instrum. **3**(08):S08004 (2008)
7. Aamodt, K., et al.: The ALICE Collaboration, The ALICE experiment at the CERN LHC. J. Instrum. **3**(08):S08002 (2008)
8. Aad, G., et al.: The ATLAS Collaboration, The ATLAS experiment at the CERN large hadron collider. J. Instrum. **3**(08):S08003 (2008)
9. Spira, M., Djouadi, A., Graudenz, D., Zerwas, R.M.: Higgs boson production at the LHC. Nucl. Phys. B **453**(1–2), 17–82 (1995)
10. Hinchliffe, I., Paige, F.E., Shapiro, M.D., Söderqvist, J., Yao, W.: Precision SUSY measurements at CERN LHC. Phys. Rev. D **55**(9), 5520–5540 (1997)
11. Dimopoulos, S., Landsberg, G.: Black holes at the large Hadron collider. Phys. Rev. Lett. **87**(16), 161602 (2001)
12. Tajima, T. (ed.): The future of accelerator physics. AIP Conference Proceedings 356. AIP (1994)
13. http://www.hep.net/ssc/
14. Mervis, J., Seife, C.: 10 years after the SSC: Lots of reasons, but few lessons. Science **302**, 38–40 (2003)
15. Mervis, J.: 10 years after the SSC: scientists are long gone, but bitter memories remain. Science **302**, 40–41 (2003)
16. http://www.linearcollider.org/
17. Barish, B.: Ilc/gde report. Proceedings of TILC 09, (2009)
18. Wu Chao, A., Tigner, M.: Handbook of Accelerator Physics and Engineering. World Scientific Publishing Co Pte Ltd (1999)
19. Koch, E.E.: Particle Accelerator Physics. 3rd edn. Springer, Berlin (2007)
20. Tajima, T., Dawson, J.M.: Laser electron accelerator. Phys. Rev. Lett. **43**(4), 267 (1979)
21. Rosenzweig, J.B., Cline, D.B., Cole, B., Figueroa, H., Gai, W., Konecny, R., Norem, J., Schoessow, P., Simpson, J.: Experimental observation of plasma wake-field acceleration. Phys. Rev. Lett. **61**(1), 98 (1988)
22. Rosenzweig, J.B., Schoessow, P., Cole, B., Gai, W., Konecny, R., Norem, J., Simpson, J.: Experimental measurement of nonlinear plasma wake fields. Phys. Rev. A **39**(3), 1586–1589 (1989)
23. Nakanishi, H., Enomoto, A., Ogata, A., Nakajima, K., Whittum, D., Yoshida, Y., Ueda, T., Kobayashi, T., Shibata, H., Tagawa, S., Yugami, N., Nishida, Y.: Wakefield accelerator using twin linacs. Nucl. Instrum. Methods Phys. Res. Sect. A: Accel. Spectrom. Detectors Assoc. Equip. **328**(3), 596–598 (1993)
24. Berezin, A.K., Fainberg, Ya.B., Kiselev, V.A., Linnik, A.F., Uskov, V.V., Balakirev, V.A., Onishchendo, I.N., Sidelnikov, G.L., Sotnikov, G.V.: Wake field excitation in plasma by a relativistic electron pulse with a controlled number of short bunches. Plasma Phys. Rep. **20**, 596 (1994)
25. Hogan, M.J., Barnes, C.D., Clayton, F.J., Decker, C.E., Deng, S., Emma, P., Huang, C., Iverson, R.H., Johnson, D.K., Joshi, C., Katsouleas, T., Krejcik, P., Lu, W., Marsh, K.A.,

Mori, W.B., Muggli, P., O'Connell, C.L., Oz, E., Siemann, R.H., Walz, D.: Multi-gev energy gain in a plasma-wakefield accelerator. Phys. Rev. Lett. **95**, 054802 (2005)
26. Blumenfeld, I., Clayton, C.E., Decker, F.-J., Hogan, M.J., Huang, C., Ischebeck, R., Iverson, R., Joshi, C., Katsouleas, T., Kirby, N., Lu, W., Marsh, K.A., Mori, W.B., Muggli, P., Oz, E., Siemann, R.H., Walz, D., Zhou, M.: Energy doubling of 42 GeV electrons in a metre-scale plasma wakefield accelerator. Nature **445**(7129), 741–744 (2007)
27. Kitagawa, Y., Matsumoto, T., Minamihata, T., Sawai, K., Matsuo, K., Mima, K., Nishihara, K., Azechi, H., Tanaka, K.A., Takabe, H., Nakai, S.: Beat-wave excitation of plasma wave and observation of accelerated electrons. Phys. Rev. Lett. **68**(1), 48–51 (1992)
28. Clayton, C.E., Everett, M.J., Lal, A., Gordon, D., Marsh, K.A., Joshi, C.: Acceleration and scattering of injected electrons in plasma beat wave accelerator experiments. Phys. Plasmas **1**(5), 1753–1760 (1994)
29. Everett, M., Lal, A., Gordon, D., Clayton, C.E., Marsh, K.A., Joshi, C.: Trapped electron acceleration by a laser-driven relativistic plasma wave. Nature **368**(6471), 527–529 (1994)
30. Ebrahim, N.A.: Optical mixing of laser light in a plasma and electron acceleration by relativistic electron plasma waves. J. Appl. Phys. **76**(11), 7645–7647 (1994)
31. Amiranoff, F., Ardonceau, J., Bercher, M., Bernard, D., Cros, B., Debraine, A., Dieulot, J.M., Fusellier, J., Jacquet, F., Joly, J.M., Juillard, M., Matthieussent, G., Matricon, P., Mine, P., Montes, B., Mora, P., Morano, R., Morillo, J., Moulin, F., Poilleux, P., Specka, A., Stenz, C.: Electron acceleration in the plasma beat-wave experiment at ecole polytechnique. In : Proceedings of the AIP Conference on Advanced Accelerator Concepts, **335**, 612 (1995)
32. Hamster, H., Sullivan, A., Gordon, S., White, W., Falcone, R.W.: Subpicosecond, electromagnetic pulses from intense laser-plasma interaction. Phys. Rev. Lett. **71**(17), 2725–2728 (1993)
33. Nakajima, K., Kawakubo, T., Nakanishi, H., Ogata, A., Kitagawa, Y., Kodama, R., Mima, K., Shiraga, H., Suzuki, K., Yamakawa, K., Zhang, T., Kato, Y., Fisher, D., Downer, M., Tajima, T., Sakawa, Y., Shoji, T., Yugami, N., Nishida, N.: Proof-ofprinciple experiments of laser wakefield acceleration using a 1 ps 10 TW Nd:glass laser. In: Proceedings of the AIP Conference on Advanced Accelerator Concepts, pp. 145–155 (1995)
34. Downer, M.C., Siders, C.W., Fisher, D.F., LeBlanc, S.P., Rau, B., Gaul, E., Tajima, T., Babine, A., Stepanov, A., Sergeev, A.: Laser wakefield photon accelerator: optical diagnostics for the laser wakefield accelerator based on longitudinal interferometry. Bullet Am. Phys. Soc. **40**, 1862 (1995)
35. Marquès, J.R., Geindre, J.P., Amiranoff, F., Audebert, P., Gauthier, J.C., Antonetti, A., Grillon, G.: Temporal and spatial measurements of the electron density perturbation produced in the wake of an ultrashort laser pulse. Phys. Rev. Lett. **76**(19), 3566–3569 (1996)
36. Gorbunov, L.M., Kirsanov, V.I.: Excitation of plasma waves by an electromagnetic wave packet. SOV Phys. JETP **66**, 290–294 (1987)
37. Bulanov, S.V., Kirsanov, V.I., Sakharov, A.S.: Excitation of ultrarelativistic plasma waves by pulse of electromagnetic radiation. JETP Lett. **50**, 198–201 (1989)
38. Sprangle, P., Esarey, E., Ting, A.: Nonlinear theory of intense laser-plasma interactions. Phys. Rev. Lett. **64**(17), 2011–2014 (1990)
39. Sprangle, P., Esarey, E., Ting, A.: Nonlinear interaction of intense laser pulses in plasmas. Phys. Rev. A **41**, 4463–4469 (1990)
40. Coverdale, C.A., Darrow, C.B., Decker, C.D., Mori, W.B., Tzeng, K.-C., Marsh, K.A., Clayton, C.E., Joshi, C.: Propagation of intense subpicosecond laser pulses through underdense plasmas. Phys. Rev. Lett. **74**(23), 4659–4662 (1995)
41. Nakajima, K., Fisher, D., Kawakubo, T., Nakanishi, H., Ogata, A., Kato, Y., Kitagawa, Y., Kodama, R., Mima, K., Shiraga, H., Suzuki, K., Yamakawa, K., Zhang, T., Sakawa, Y., Shoji, T., Nishida, Y., Yugami, N., Downer, M., Tajima, T.: Observation of ultrahigh gradient electron acceleration by a self-modulated intense short laser pulse. Phys. Rev. Lett. **74**(22), 4428–4431 (1995)

42. Modena, A., Najmudin, Z., Dangor, A.E., Clayton, C.E., Marsh, K.A., Joshi, C., Malka, V., Darrow, C.B., Danson, C., Neely, D., Walsh, F.N.: Electron acceleration from the breaking of relativistic plasma waves. Nature **377**(6550), 606–608 (1995)
43. Wagner, R., Chen, S.-Y., Maksimchuk, A., Umstadter, D.: Electron acceleration by a laser wakefield in a relativistically self-guided channel. Phys. Rev. Lett. **78**(16), 3125–3128 (1997)
44. Moore, C.I., Ting, A., Krushelnick, K., Esarey, E., Hubbard, R.F., Hafizi, B., Burris, H.R., Manka, C., Sprangle, P.: Electron trapping in self-modulated laser wakefields by raman backscatter. Phys. Rev. Lett. **79**(20), 3909–3912 (1997)
45. Ting, A., Moore, C.I., Krushelnick, K., Manka, C., Esarey, E., Sprangle, P., Hbbard, R., Burris, H.R., Fischer, R., Baine, M.: Plasma wakefield generation and electron acceleration in a self-modulated laser wakefield accelerator experiment. Phys. Plasmas **4**(5), 1889–1899 (1997)
46. Santala, M.I.K., Najmudin, Z., Clark, E.L., Tatarakis, M., Krushelnick, K., Dangor, A.E., Malka, V., Faure, J., Allott, R., Clarke, R.J.: Observation of a hot highcurrent electron beam from a self-modulated laser wakefield accelerator. Phys. Rev. Lett. **86**(7), 1227–1230 (2001)
47. Malka, V., Fritzler, S., Lefebvre, E., Aleonard, M.-M., Burgy, F., Chambaret, J.-P., Chemin, J.-F., Krushelnick, K., Malka, G., Mangles, S.P.D., Najmudin, Z., Pittman, M., Rousseau, J.-P., Scheurer, J.-N., Walton, B., Dangor, A.E.: Electron acceleration by a wake field forced by an intense ultrashort laser pulse. Science **298**(5598), 1596–1600 (2002)
48. Esarey, E., Sprangle, P., Krall, J., Ting, A.: Overview of plasma-based accelerator concepts. IEEE Trans. Plasma Sci. **24**(2), 252 (1996)
49. Strickland, D., Mourou, G.: Compression of amplified chirped optical pulses. Opt. Commun. **56**(3), 219 (1985)
50. Andreev, N.E., Gorbunov, L.M., Kirsanov, V.I., Pogosova, A.A., Ramazashvili, R.R.: Resonant excitation of wakefields by a laser pulse in a plasma. JETP Lett. **55**, 571–576 (1992)
51. Antonsen, T.M., Mora, P.: Self-focusing and raman scattering of laser pulses in tenuous plasmas. Phys. Rev. Lett. **69**(15), 2204–2207 (1992)
52. Esarey, E., Sprangle, P., Krall, J., Ting, A., Joyce, G.: Optically guided laser wakefield acceleration. Phys. Fluids B: Plasma Phys. **5**(7), 2690–2697 (1993)
53. Litvak, A.G.: Finite-amplitude wave beams in a magnetoactive plasma. Sov Phys. JETP **30**, 344 (1970)
54. Max, C.E., Arons, J., Langdon, A.B.: Self-modulation and self-focusing of electromagnetic waves in plasmas. Phys. Rev. Lett. **33**(4), 209–212 (1974)
55. Tajima, T.: High energy laser plasma accelerators. Laser Part Beam **3**(4), 351–413 (1985)
56. Barnes, D.C., Kurki-Suonio, T., Tajima, T.: Laser self-trapping for the plasma fiber accelerator. IEEE Trans. Plasma Sci. **15**(2), 154–160 (1987)
57. Sprangle, P., Esarey, E., Ting, A., Joyce, G.: Laser wakefield acceleration and relativistic optical guiding. Appl. Phys. Lett. **53**(22), 2146–2148 (1988)
58. Forslund, D.W., Kindel, J.M., Lindman, E.L.: Theory of stimulated scattering processes in laser-irradiated plasmas. Phys. Fluids **18**(8), 1002–1016 (1975)
59. Mori, W.B., Decker, C.D., Hinkel, D.E., Katsouleas, T.: Raman forward scattering of short-pulse high-intensity lasers. Phys. Rev. Lett. **72**(10), 1482–1485 (1994)
60. Gahn, C., Tsakiris, G.D., Pukhov, A., Meyer-ter Vehn, J., Pretzler, G., Thirolf, P., Habs, D., Witte, K.J.: Multi-mev electron beam generation by direct laser acceleration in high-density plasma channels. Phys. Rev. Lett. **83**(23), 4772–4775 (1999)
61. Geissler, M., Schreiber, J., Meyer-Ter-Vehn, J.: Bubble acceleration of electrons with few-cycle laser pulses. New J. Phys. **8**, 186 (2006)
62. Pukhov, A., Meyer-Ter-Vehn, J.: Laser wake field acceleration: the highly nonlinear broken-wave regime. Appl. Phys. B **74**, 355 (2002)
63. Lu, W., Huang, C., Zhou, M., Mori, W.B., Katsouleas, T.: Nonlinear theory for relativistic plasma wakefields in the blowout regime. Phys. Rev. Lett. **96**(16), 165002 (2006)

64. Tsung, F.S., Lu, W., Tzoufras, M., Mori, W.B., Joshi, C., Vieira, J.M., Silva, L.O., Fonseca, R.A.: Simulation of monoenergetic electron generation via laser wakefield accelerators for 5–25 TW lasers. Phys. Plasmas **13**(5), 056708 (2006)
65. Gordienko, S., Pukhov, A.: Scalings for ultrarelativistic laser plasmas and quasimonoenergetic electrons. Phys. Plasmas **12**, 043109 (2005)
66. Pukhov, A., Gordienko, S.: Bubble regime of wake field acceleration: similarity theory and optimal scalings. Phil. Trans. R. Soc. A **364**, 623 (2006)
67. Hafz, N.A.M., Jeong, T.M., Choi, I.W., Lee, S.K., Pae, K.H., Kulagin, V.K., Sung, J.H., Yu, T.J., Hong, K.-H., Hosokai, T., Cary, J.R., Ko, D.-K., Lee, J.: Stable generation of GeV-class electron beams from self-guided laser-plasma channels. Nat. Phot. **2**, 571 (2008)
68. Faure, J., Glinec, Y., Pukhov, A., Kiselev, S., Gordienko, S., Lefebvre, E., Rousseau, J.-P., Burgy F., Malka, V. : A laser-plasma accelerator producing monoenergetic electron beams. Nature **431**, 541 (2004)
69. Geddes, C.G.R. et al.: High quality electron beams from a laser wakefield accelerator using plasma-channel guiding. Nature **431**, 538 (2004)
70. Mangles, S.P.D., Murphy, C.D., Najmudin, Z., Thomas, A.G.R., Collier, J.L., Dangor, A.E., Divall, E.J., Foster, P.S., Gallacher, J.G., Hooker, C.J., Jaroszynski, D.A., Langley, A.J., Mori, W.B., Norreys, P.A., Tsung, F.S., Viskup, R., Walton, B.R., Krushelnick, K.: Monoenergetic beams of relativistic electrons from intense laser plasma interactions. Nature **431**, 535 (2004)
71. Hidding, B., Amthor, K.-U., Liesfeld, B., Schwoerer, H., Karsch, S., Geissler, M., Veisz, L., Schmid, K., Gallacher, J.G., Jamison, S.P., Jaroszynski, D., Pretzler, G., Sauerbrey, R.: Generation of quasimonoenergetic electron bunches with 80-fs laser pulses. Phys. Rev. Lett. **96**(10), 105004 (2006)
72. Leemans, W.P., Nagler, B., Gonsalves, A.J., Toth C, s., Nakamura, K., Geddes, C.G.R., Esarey, E., Schroeder, C.B., Hooker, S.M.: Gev electron beams from a centimetre-scale accelerator. Nature Phys. **2**, 696 (2006)
73. Faure, J., Rechatin, C., Norlin, A., Lifschitz, A., Glinec, Y., Malka, V.: Controlled injection and acceleration of electrons in plasma wakefields by colliding laser pulses. Nature **444**, 737 (2006)
74. Rechatin, C., Faure, J., Ben-Ismail, A., Lim, J., Fitour, R., Specka, A., Videau, H., Tafzi, A., Burgy, F., Malka, V.: Controlling the phase-space volume of injected electrons in a laser-plasma accelerator. Phys. Rev. Lett. **102**(16), 164801 (2009)
75. Osterhoff, J., Popp, A., Major Z, s., Marx, B., Rowlands-Rees, T.P., Fuchs, M., Geissler, M., Hörlein, R., Hidding, B., Becker, S., Peralta, E.A., Schramm, U., Grüner, F., Habs, D., Krausz, F., Hooker, S.M., Karsch, S.: Generation of stable, low-divergence electron beams by laser-wakefield acceleration in a steady-stateflow gas cell. Phys. Rev. Lett. **101**(8), 085002 (2008)
76. Tavella, F., Nomura, Y., Veisz, L., Pervak, V., Marcinkevičius, A., Krausz, F.: Dispersion management for a sub-10-fs, 10 TW optical parametric chirped-pulse amplifier. Opt. Lett. **32**(15), 2227–2229 (2007)
77. Mangles, S.P.D., Walton, B.R., Tzoufras, M., Najmudin, Z., Clarke, R.J., Dangor, A.E., Evans, R.G., Fritzler, S., Gopal, A., Hernandez-Gomez, C., Mori, W.B., Rozmus, W., Tatarakis, M., Thomas, A.G.R., Tsung, F.S., Wei, M.S., Krushelnick, K.: Electron acceleration in cavitated channels formed by a petawatt laser in low-density plasma. Phys. Rev. Lett. **94**(24), 245001 (2005)
78. Malka, V., Faure, J., Glinec, Y., Pukhov, A., Rousseau, J.-P.: Monoenergetic electron beam optimization in the bubble regime. Phys. Plasmas **12**(5), 056702 (2005)
79. Geddes, C.G.R., Toth, Cs., van Tilborg, J., Esarey, E., Schroeder, C.B., Bruhwiler, D., Nieter, C., Cary, J., Leemans, W.P.: Production of high-quality electron bunches by dephasing and beam loading in channeled and unchanneled laser plasma accelerators. Phys. Plasmas **12**, 056709 (2005)
80. Hosokai, T., Kinoshita, T., Ohkubo, T., Maekawa, A., Uesaka, M., Zhidkov, A., Yamazaki, A., Kotaki, H., Kando, M., Nakajima, K., Bulanov, S.V., Tomassini, P., Giulietti, A., Giulietti, D.:

Observation of strong correlation between quasimonoenergetic electron beam generation by laser wakefield and laser guiding inside a preplasma cavity. Phys. Rev. E **73**(3), 036407 (2006)

81. Maksimchuk, A., Reed, S., Bulanov, S.S., Chvykov, V., Kalintchenko, G., Matsuoka, T., McGuffey, C., Mourou, G., Naumova, N., Nees, J., Rousseau, P., Yanovsky, V., Krushelnick, K., Matlis, N.H., Kalmykov, S., Shvets, G., Downer, M.C., Vane, C.R., Beene, J.R., Stracener, D., Schultz, D.R.: Studies of laser wakefield structures and electron acceleration in underdense plasmas. Phys. Plasmas **15**(5), 056703 (2008)
82. Yamazaki, A., Kotaki, H., Daito, I., Kando, M., Bulanov, S.V., Esirkepov, T.Zh., Kondo, S., Kanazawa, S., Homma, T., Nakajima, K., Oishi, Y., Nayuki, T., Fujii, T., Nemoto, K.: Quasi-monoenergetic electron beam generation during laser pulse interaction with very low density plasmas. Phys. Plasmas **12**(9), 093101 (2005)
83. Masuda, S., Miura, E., Koyama, K., Kato, S., Adachi, M., Watanabe, T., Torii, K., Tanimoto, M.: Energy scaling of monoenergetic electron beams generated by the laser-driven plasma based accelerator. Phys. Plasmas **14**(2), 023103 (2007)
84. Miura, E., Koyama, K., Kato, S., Saito, N., Adachi, M., Kawada, Y., Nakamura, T., Tanimoto, M.: Demonstration of quasi-monoenergetic electron-beam generation in laser-driven plasma acceleration. Appl. Phys. Lett. **86**, 251501 (2005)
85. Hsieh, C.-T., Huang, C.-M., Chang, C.-L., Ho, Y.-C., Chen, Y.-S., Lin, J.-Y., Wang, J., Chen, S.-Y.: Tomography of injection and acceleration of monoenergetic electrons in a laser-wakefield accelerator. Phys. Rev. Lett. **96**(9), 095001 (2006)
86. Leemans, W.P., Geddes, C.G.R., Faure, J., Tóth, Cs., van Tilborg, J., Schroeder, C.B., Esarey, E., Fubiani, G., Auerbach, D., Marcelis, B., Carnahan, M.A., Kaindl, R.A., Byrd, J., Martin, M.C.: of terahertz emission from a laser-plasma accelerated electron bunch crossing a plasma-vacuum boundary. Phys. Rev. Lett. **91**(7), 074802 (2003)
87. Schroeder, C.B., Esarey, E., van Tilborg, J., Leemans, W.P.: Theory of coherent transition radiation generated at a plasma-vacuum interface. Phys. Rev. E **69**(016501), (2004)
88. van Tilborg, J., Schroeder, C.B., Filip, C.V., Tóth, Cs., Geddes, C.G.R., Fubiani, G., Huber, R., Kaindl, R.A., Esarey, E., Leemans, W.P.: Temporal characterization of femtosecond laser-plasma-accelerated electron bunches using terahertz radiation. Phys. Rev. Lett. **96**(1), 014801 (2006)
89. Butler, A., Gonsalves, A.J., McKenna, C.M., Spence, D.J., Hooker, S.M., Sebban, S., Mocek, T., Bettaibi, I., Cros, B.: Demonstration of a collisionally excited optical-field-ionization XUV laser driven in a plasma waveguide. Phys. Rev. Lett. **91**(20), 205001 (2003)
90. Rousse, A., Phuoc, K.T., Shah, R., Pukhov, A., Lefebvre, E., Malka, V., Kiselev, S., Burgy, F., Rousseau, J.-P., Umstadter, D., Hulin, D.: Production of a KeV X-ray beam from synchrotron radiation in relativistic laser-plasma interaction. Phys. Rev. Lett. **93**(13), 135005 (2004)
91. Phuoc, K.T., Burgy, F., Rousseau, J.-P., Malka, V., Rousse, A., Shah, R., Umstadter, D., Pukhov, A., Kiselev, S.: Laser based synchrotron radiation. Phys. Plasmas, **12**(2), 023101 (2005)
92. Phuoc, K.T., Corde, S., Shah, R., Albert, F., Fitour, R., Rousseau, J.-P., Burgy, F., Mercier, B., Rousse, A.: Imaging electron trajectories in a laser-wakefield cavity using betatron X-ray radiation. Phys. Rev. Lett. **97**(22), 225002 (2006)
93. Kneip, S., Nagel, S.R., Bellei, C., Bourgeois, N., Dangor, A.E., Gopal, A., Heathcote, R., Mangles, S.P.D., Marquès, J.R., Maksimchuk, A., Nilson, P.M., Phuoc, K.Ta., Reed, S., Tzoufras, M., Tsung, F.S., Willingale, L., Mori, W.B., Rousse, A., Krushelnick, K., Najmudin, Z.: Observation of synchrotron radiation from electrons accelerated in a Petawatt-laser-generated plasma cavity. Phys. Rev. Lett. **100**(10), 105006 (2008)
94. Albert, F., Shah, R., Phuoc, K.T., Fitour, R., Burgy, F., Rousseau, J.-P., Tafzi, A., Douillet, D., Lefrou, T., Rousse, A.: Betatron oscillations of electrons accelerated in laser wakefields characterized by spectral X-ray analysis. Phys. Rev. E **77**(5), 056402 (2008)
95. Krushelnick, K., Clark, E.L., Najmudin, Z., Salvati, M., Santala, M.I.K., Tatarakis, M., Dangor, A.E., Malka, V., Neely, D., Allott, R., Danson, C.: Multimev ion production from

high-intensity laser interactions with underdense plasmas. Phys. Rev. Lett. **83**(4), 737–740 (1999)
96. Willingale, L., Mangles, S.P.D, Nilson, P.M, Clarke, R.J, Dangor, A.E, Kaluza, M.C, Karsch, S., Lancaster, K.L, Mori, W.B, Najmudin, Z., Schreiber, J., Thomas, A.G.R, Wei, M.S, Krushelnick, K.: Collimated multi-mev ion beams from high-intensity laser interactions with underdense plasma. Phys. Rev. Lett. **96**(24), 245002 (2006)
97. Faure, J., Glinec, Y., Santos, J.J., Ewald, F., Rousseau, J.-P., Kiselev, S., Pukhov, A., Hosokai, T., Malka, V.: Observation of laser-pulse shortening in nonlinear plasma waves. Phys. Rev. Lett. **95**(20), 205003 (2005)
98. Kando, M., Fukuda, Y., Pirozhkov, A.S., Ma, J., Daito, I., Chen, L.-M., Esirkepov, T.Zh., Ogura, K., Homma, T., Hayashi, Y., Kotaki, H., Sagisaka, A., Mori, M., Koga, J.K., Daido, H., Bulanov, S.V., Kimura, T., Kato, Y., Tajima, T.: Demonstration of laser-frequency upshift by electron-density modulations in a plasma wakefield. Phys. Rev. Lett. **99**(13), 135001 (2007)
99. Pegoraro, F., Bulanov, S.V., Califano, F., Esirkepov, T.Zh., Lontano, M., Meyer-ter Vehn, J., Naumova, N.M., Pukhov, A.M., Vshivkov, V.A.: Magnetic fields from high-intensity laser pulses in plasmas. Plasma Phys. Control Fusion **39**, 261–272 (1997)
100. Bulanov, S.S., Esirkepov, T.Zh., Kamenets, F.F., Pegoraro, F.: Single-cycle highintensity electromagnetic pulse generation in the interaction of a plasma wakefieldwith regular nonlinear structures. Phys. Rev. E (Stat, Nonlinear, Soft Matter Phys.) **73**(3), 036408 (2006)
101. Dun, H., Mattes, B.L., Stevenson, D.A.: The gas dynamics of a conical nozzle molecular beam sampling system. Chem. Phys. **38**, 161 (1979)
102. Knuth, E.L.: Size correlations for condensation clusters produced in free-jet expansions. J. Chem. Phys. **107**(21), 9125–9132 (1997)
103. Ditmire, T., Smith, R.A.: Short-pulse laser interferometric measurement of absolute gas densities from a cooled gas jet. Opt. Lett. **23**(8), 618 (1998)
104. Smith, R.A., Ditmire, T., Tisch, J.W.G.: Characterization of a cryogenically cooled high-pressure gas laser/cluster interaction experiments. Rev. Sci. Inst. **69**(11), 3798 (1998)
105. Khoukaz, A., Lister, T., Quentmeier, C., Santo, R., Thomas, C.: Systematic studies on hydrogen cluster beam production. Eur. Phys. J. D **5**, 275 (1999)
106. Pedemonte, L., Bracco, G., Tatarek, R.: Theoretical and experimental study of he free-jet expansions. Phys. Rev. A **59**(4), 3084 (1999)
107. Even, U., Jortner, J., Noy, D., Lavie, N., Cossart-Magos, C.: Cooling of large molecules below 1 k and He clusters formation. J. Chem. Phys. **112**(18), 8068 (2000)
108. Parra, E., McNaught, S.J., Milchberg, H.M.: Characterization of a cryogenic, high-pressure gas jet operated in the droplet regime. Rev. Sci. Instrum. **73**(2), 468–475 (2002)
109. Kim, K.Y., Kumarappan, V., Milchberg, H.M.: Measurement of the average size and density of clusters in a gas jet. Appl. Phys. Lett. **83**(15), 3210–3212 (2003)
110. Lawrence, L.S., French, R.J.: Electron diffraction investigation of pulsed supersonic jets. Rev. Sci. Inst. **60**(7), 1223 (1989)
111. Pronko, J., Kohler, D., Chapman, I.V., Bardin, T.T., Filbert, P.C., Hawley, J.D.: Density measurement of a pulsed supersonic gas jet using nuclear scattering. Rev. Sci. Inst. **64**, 1744 (1993)
112. Perry, M.D., Darrow, C., Coverdale, C., Crane, J.K.: Measurement of the local electron density by means of stimulated raman scattering in a laser-produced gas jet plasma. Opt. Lett. **17**(7), 523 (1992)
113. Lompré, L.A., Ferray, M., L'Huillier, A., Li, X.F., Mainfray, G.: Optical determination of the characteristics of a pulsed gas jet. J. Appl. Phys. **63**(5), 1791 (1988)
114. Tejeda, G., Maté, B., Fernández-Sánchez, J.M., Montero, S.: Temperature and density mapping of supersonic jet expansions using linear raman spectroscopy. Phys. Rev. Lett. **76**(1), 34–37 (1996)
115. Winckler, J.: The mach interferometer applied to studying an axially symmetric supersonic air jet. Rev. Sci. Instrum. **19**(5), 307–322 (1948)

116. Behjat, A., Tallents, G.J., Neely, D.: The characterization of a high-density gas jet. J. Phys. D: Appl. Phys. **30**, 2872 (1997)
117. Auguste, T., Bougeard, M., Caprin, E., D'Oliveira, P., Monot, P.: Characterization of a high-density large scale pulsed gas jet for laser–gas interaction experiments. Rev. Sci. Instrum. **70**(5), 2349–2354 (1999)
118. Azambuja†, R., Eloy, M., Figueira, G., Neely, D.: Three-dimensional characterization of high-density non-cylindrical pulsed gas jets. J. Phys. D: Appl. Phys. **32**, 35 (1999)
119. Malka, V., Coulaud, C., Geindre, J.P., Lopez, V., Najmudin, Z., Neely, D., Amiranoff, F.: Characterization of neutral density profile in a wide range of pressure of cylindrical pulsed gas jets. Rev. Sci. Instrum. **71**(6), 2329–2333 (2000)
120. Kim, C., Kim, G.-H., Kim, J.-U., Ko, I.S., Suk, H.: Characterizations of symmetry and asymmetry high-density gas jets without abel inversion. Rev. Sci. Instrum. **75**(9), 2865–2868 (2004)
121. Semushin, S., Malka, V.: High density gas jet nozzle design for laser target production. Rev. Sci. Inst. **72**(7), 2961 (2001)
122. Hosokai T., et al. Supersonic gas jet target for generation of relativistic electrons with 12 TW-50 fs laser pulse. In: Proceedings of EPAC 2002, pp. 981–983, (2002)
123. Janson S.W., Helvajian H., Breuer K. Mems, Microengineering and aerospace systems. AIAA (99–3802), (1999)
124. Hitt, D.L., Zakrzwski, C.M., Thomas, M.A.: Mems-based satellite micropropulsion via catalyzed hydrogen peroxide decomposition. Smart Mater. Struct. **10**, 1163 (2001)
125. Xie, C.: Characteristics of micronozzle gas flows. Phys. Fluids **19**(3), 037102 (2007)
126. Broc, A., de Benedictis, S., Dilecce, G., Vigliotti, M., Sharafutdinov, R.G., Skovorodko, P.A.: Experimental and numerical investigation of an O2/NO supersonic free jet expansion. J. Fluid Mech. **500**, 211 (2004)
127. Boyd, I.D., Beattie, D.R., Cappelli, M.A.: Numerical and experimental investigations of low-density supersonic jets of hydrogen. J. Fluid Mech. **280**, 41 (1994)
128. Boyd, I.D., Chen, G., Candler, G.: Predicting failure of the continuum fluid equations in translational hypersonic flows. Phys. Fluids **7**(1), 210 (1995)
129. Mo H., Lin C., Gokaltun S., Skudarnov P.V.: Numerical study of axisymmetric gas flow in conical micronozzles by DSMC an continuum methods. AIAA (2006–991) (2006)
130. Agarwal, R.K.: Beyond Navier-Stokes: Burnett equations for flows in the continuum-transition regime. Phys. Fluids **13**(10), 3061–3085 (2001)
131. Pandey, B.P., Raju, R., Roy, S., Finite element model of fluid flow inside a micro thruster. AIAA (2002–5733),(2002)
132. Gadepalli, V.V.V., Lin, C. NavierSstokes modeling of gas flows in a de-laval micronozzle. AIAA (2006–1425), (2006)
133. Hao, P.-F., Ding, Y.-T., Yao, Z.-H., He, F., Zhu, K.-Q.: Size effect on gas flow in micro nozzles. J. Micromech. Microeng. **15**, 2069 (2005)
134. Alexeenko, A.A., Fedosov, D.A., Gimelshein, S.F., Levin, D.A., Collins, R.J.: Transient heat transfer and gas flow in a mems-based thruster. J. Microelectromech **15**(1), 181 (2006)
135. Louisos, W.F., Hitt, D.L.: Optimal expansion angle for viscous supersonic flow in 2-d micro nozzles. AIAA, (2005–5032), (2005)
136. Alexeenko, A.A., Levin, D.A., Gimelshein, S.F., Collins, R.J., Reed, B.D.: Numerical modeling of axisymmetric and three-dimensional flows in microelectromechanical systems nozzles. AIAA J. **40**(5), 897 (2002)
137. Mate, B., Graur, I.A., Elizarova, T., Chirokov, I., Tejeda, G., Fernandez, J.M., Montero, S.: Experimental and numerical investigation of an axisymmetric supersonic jet. J. Fluid Mech. **426**, 177 (2001)
138. Ketsdever, A., Wadsworth, D.C., Wapner, P.G., Ivanov, M.S., Markelov, G.N.: Fabrication and predicted performance of conical delaval micronozzles. AIAA (99–2724) (1999)

Part I
Supersonic Micro-Jets

Chapter 2
Theory of Compressible Fluid Flow

2.1 One-Dimensional Theory of Compressible Fluid Flow

2.1.1 Equation of State and the First Principal Law

The thermodynamic properties of a homogeneous and isotropic medium are fully characterized by the three quantities *temperature* T (K), *pressure* p (Pa), and *density* $\rho\,(\mathrm{kg/m^3})$ which are called the state variables. About the microscopic structure of the medium the only necessary assumptions are that the particles that constitute the medium are small enough to justify the assumption of homogeneity and that for any given set of T, p, ρ the composition of the medium is fixed. This includes, for example, mixtures of gases or air that contains a certain amount of water in the form of small droplets.

The quantities T, p, ρ depend on each other, their relation is given by the equation of state of that medium.

$$p = p(\rho, T) \tag{2.1}$$

Additional state variables can and will be defined and used but—together with the equation of state—always two of them are enough to characterize the state of the medium. In the case of a perfect gas the equation of state has the well-known form

$$p = \frac{R}{m}\rho T \tag{2.2}$$

where R is the gas constant that is given by $R = k_B N_A = 8314.46\,\mathrm{J/(kg\ K)}$. Here k_B is the Bolzmann constant and N_A the Avogadro constant. m is the atomic (molecular) weight of the medium.

By introducing now the general physical principle of energy conservation one immediately arrives at the first principle law of thermodynamics that essentially states the conservation of energy in its transformation from heat to mechanical

K. Schmid, *Laser Wakefield Electron Acceleration*, Springer Theses,
DOI: 10.1007/978-3-642-19950-9_2, © Springer-Verlag Berlin Heidelberg 2011

energy and vice versa. The quantity that measures the energy content of a medium is the inner energy e and is a state variable on its own. Hence, it only depends on two other state variables, for example, $e = e(p, \rho)$ which again represents an equation of state. In the case of the perfect gas, the inner energy only depends on the temperature $e = e(T)$. This inner energy is constant as long as the gas is not doing any work thereby converting a part of its inner energy into mechanical energy. This does not entirely exclude changes of the volume of the gas as can be seen in the following case: a perfectly isolated vessel that is separated in two compartments by a removable wall with one compartment being filled with a gas of a certain pressure and temperature. By removing the wall, the gas will now expand and fill the entire vessel. Since there is now energy (heat) transport through the vessel walls, the inner energy must be conserved which means for a perfect gas that after the expansion is completed the gas will still have the initial temperature. This is a so-called irreversible adiabatic expansion because it is impossible to reach the initial state without energy transfer to the gas. If the wall is now replaced by a slowly moving piston, the situation is different: the gas exerts a force on the piston corresponding to its pressure times the piston surface. By moving the piston a distance ds, the work $pAds$ is done. Since Ads is equal to the change of the volume of the gas, the amount of work done is given by $pdV = p\mathrm{d}(1/\rho)$. If heat transport to the medium dq is now considered, then the law of energy conservation takes on the form:

$$\mathrm{d}q = \mathrm{d}e + p\mathrm{d}(1/\rho) \tag{2.3}$$

Here the amount of heat q is measured in joule per kg. This equation is called the first principle law of thermodynamics. With the enthalpy i being defined as

$$i = e + p/\rho \tag{2.4}$$

it can be written as

$$\mathrm{d}q = \mathrm{d}i - 1/\rho\mathrm{d}p \tag{2.5}$$

Heat can be transferred to a medium in different ways, the two most easily realizable methods are maintaining the volume and hence the density constant and keeping the pressure constant. By measuring the amount of energy needed to raise the temperature of the medium by 1 K one arrives at the definition of the specific heats c_p and c_v. They are defined as

$$c_v = \left(\frac{\partial q}{\partial T}\right)_v = \left(\frac{\partial e}{\partial T}\right)_v \tag{2.6}$$

for constant volume and as

$$c_p = \left(\frac{\partial q}{\partial T}\right)_p = \left(\frac{\partial i}{\partial T}\right)_p \tag{2.7}$$

for constant pressure. The ratio between the specific heat is ascribed the symbol κ and is defined as

$$\kappa = c_p/c_v \quad \kappa = \frac{2+f}{f} \tag{2.8}$$

The equation to the right gives a connection to the kinetic theory of gases with f corresponding to the number of degrees of freedom of the molecules constituting the medium. Possible values for f are therefore three for atoms, five for molecules containing two atoms, and seven for molecules containing three or more atoms.

For the perfect gas, one obtains for the specific heats

$$\mathrm{d}e = c_v \mathrm{d}T \tag{2.9}$$

and

$$\mathrm{d}i = c_p \mathrm{d}T \tag{2.10}$$

Substituting this into (2.4) and using (2.2) we get

$$R/m = c_p - c_v \tag{2.11}$$

With the additional assumption that a medium not only behaves like a perfect gas but also has constant specific heats c_p and c_v for all temperatures from (2.9) and (2.10) we get

$$e = c_v T + const \tag{2.12}$$

and

$$i = c_p T + const \tag{2.13}$$

2.1.2 Changes of State

A medium can change its thermodynamic state in many different ways and it makes sense to distinguish between several special cases of such processes. In general, this is done by categorizing by state variables that remain constant during the whole process. This immediately gives the three processes that keep one of the state variables in the equation of state (2.1) constant, namely the isothermal, the isobaric, and the isochoric process. For many gas flows, however, another process is of interest: it is called isentropic process, is equivalent to a reversible adiabatic process, and is defined by the absence of heat flux across the boundaries of the system and also within the medium. This means that each small volume dV of the medium is in pressure-equilibrium with its surroundings and does not receive or loose any heat energy. The validity of this assumption will be discussed later.

To justify the name *isentropic*, the variable of state called entropy s is first introduced. It is defined by

$$ds = \frac{dq}{T} = \frac{de + pd(1/\rho)}{T} = \frac{di - 1/\rho dp}{T} \tag{2.14}$$

For the perfect gas with constant specific heat this can be integrated immediately and gives:

$$s_2 - s_1 = c_v ln\frac{T_2}{T_1} - (c_p - c_v) ln\frac{\rho_2}{\rho_1} = c_p ln\frac{T_2}{T_1} - (c_p - c_v) ln\frac{p_2}{p_1} = c_v ln\frac{p_2}{p_1} - c_p ln\frac{\rho_2}{\rho_1} \tag{2.15}$$

Here the subscripts 1 and 2 refer to the initial and the final state of the medium, respectively. For the isentropic process ds is equal to zero, Eq. 2.14 shows that this is equal to constant heat energy dq and for ds = 0:

$$0 = de + pd\left(\frac{1}{\rho}\right) \tag{2.16}$$

$$0 = di - \frac{1}{\rho}dp \tag{2.17}$$

with (2.9) and (2.11) this can be integrated and yields the following equations for the isentropic change of state from state 1 to state 2:

$$\frac{\rho_2}{\rho_1} = \left(\frac{T_2}{T_1}\right)^{\frac{1}{\kappa-1}} \tag{2.18}$$

$$\frac{p_2}{p_1} = \left(\frac{T_2}{T_1}\right)^{\frac{\kappa}{\kappa-1}} \tag{2.19}$$

$$\frac{p_2}{p_1} = \left(\frac{\rho_2}{\rho_1}\right)^{\kappa} \tag{2.20}$$

The concept of entropy also leads to the formulation of the second principal law of thermodynamics that states that the entropy of an isolated system can only increase or stay constant.

2.1.3 Compressible Gas Flow in 1D: Perturbations and Shocks

In the following, basic equations will be derived that describe the change of state between two spatially separated points (1) and (2) within a flowing compressible medium. In order to treat flowing media besides two thermodynamic state variables one more variable is needed to fully describe the system. This additional variable is the flow velocity w which is a scalar quantity in the one-dimensional analysis. The three equations needed to determine these variables are the continuity Eq. 2.21, the equation of motion (2.22), and the energy conservation (2.23) given here for the case of no external forces and steady state.

$$\rho_1 w_1 = \rho_2 w_2 \tag{2.21}$$

$$p_1 + \rho_1 w_1^2 = p_2 + \rho_2 w_2^2 \tag{2.22}$$

$$\frac{w_1^2}{2} + i_1 = \frac{w_2^2}{2} + i_2 \tag{2.23}$$

They describe the flow of a medium along a stream line connecting the start-point labelled by index (1), and the endpoint labelled by index (2). Since no assumption is necessary about the actual distance between the two points, the following derivation is equally valid for continuous variation between over extended distances and for discontinuous variations in the limit of zero distance. As it turns out, both cases exist in nature, the first one representing smooth gas flow without discontinuities in the state variables and the second one describing discontinuous shocks naturally emerging in supersonic flows. Both cases will be treated in the following.

Substituting now $i_2 = i_0 = i(T = 0\text{K})$ and $w_1 = 0$ in Eq. 2.23 gives the interesting result that a gas that was initially at rest and then expands freely thereby converting its whole enthalpy into kinetic energy will obtain a maximum final velocity of

$$w_{2,max} = \sqrt{2(i_1 - i_0)} \tag{2.24}$$

or, assuming a perfect gas with constant specific heat:

$$w_{2,max} = \sqrt{2c_pT} \tag{2.25}$$

For nitrogen at room temperature this evaluates to 790 m/s, for Helium to 1,765 m/s. In order to obtain one equation that describes the entire process, first the continuity Eq. 2.21 is used to write the equation of motion (2.22) as

$$\frac{w_2^2}{2} - \frac{w_1^2}{2} + \frac{1}{2}\left(\frac{1}{\rho_1} + \frac{1}{\rho_2}\right)(p_2 - p_1) = 0 \tag{2.26}$$

and then by substituting (2.23) for the velocities w_1 and w_2 finally equation

$$i_2 - i_1 = \frac{1}{2}\left(\frac{1}{\rho_1} + \frac{1}{\rho_2}\right)(p_2 - p_1) \tag{2.27}$$

is obtained. It describes a change of state along the so-called Rankine–Hugoniot curve. If the enthalpy i of the medium as a function $i(p, \rho)$ is known, then this formula describes the relation between density and pressure for the flowing medium. Another important relation, namely the one for the speed of sound, can be obtained by substituting (2.21) into (2.22), thereby eliminating w_2:

$$w_1 = \sqrt{\frac{\rho_2}{\rho_1}\frac{p_2 - p_1}{\rho_2 - \rho_1}} \tag{2.28}$$

In the limit $\Delta p \to 0$ and $\Delta \rho \to 0$ this gives the partial derivative

$$w = \sqrt{\left(\frac{\partial p}{\partial \rho}\right)_{\mathrm{H}}} \tag{2.29}$$

where the subscript H denotes the derivative to be taken along the Rankine–Hugoniot curve. This is not the speed of sound but the speed of propagation of perturbation of arbitrary strength. In fact, it will be shown below that this speed can be substantially larger than the sound-speed. For *small* pressure and density changes, however, it can be shown [1] that the change of state according to Rankine–Hugoniot and the isentropic change of state coincide up to to the second order. Therefore, for small perturbations, the well-known formula for the speed of sound is obtained.

$$w = c = \sqrt{\left(\frac{\partial p}{\partial \rho}\right)_s} \tag{2.30}$$

Here c is introduced for the sound speed and the index s indicates that the derivative has to be taken while keeping the entropy s constant. This derivation immediately shows the area of applicability of this formula: It describes the propagation speed of *small* perturbations in a compressible medium. For the perfect gas (2.30) evaluates to

$$c = \sqrt{\kappa \frac{R}{m} T} = \sqrt{c_p(\kappa - 1)T} \tag{2.31}$$

yielding at temperature of 300 K for Air 347 m/s and for Helium 1.02 km/s.

As detailed below, strong distortions are able to propagate at speeds (much) larger than the speed of sound. For the perfect gas, $i(p, \rho)$ can be obtained by substituting (2.11) and (2.8) into (2.2) and the result into (2.13):

$$i(p, \rho) = \frac{\kappa}{\kappa - 1}\frac{p}{\rho} \tag{2.32}$$

This allows to obtain from (2.27) an (implicit) relation between p and ρ along the Rankine–Hugoniot curve:

$$\frac{p_2 - p_1}{\rho_2 - \rho_1} = \kappa \frac{p_2 + p_1}{\rho_2 + \rho_1} \tag{2.33}$$

Figure 2.1a shows a plot of the density-ratio over the pressure ratio for the Rankine–Hugoniot case and the isentropic case. As can be seen, the value of the Rankine–Hugoniot curve tends asymptotically towards $(\kappa + 1)/(\kappa - 1)$ whereas the isentropic one increases with $(p_2/p_1)^{1/\kappa}$. As mentioned above, the

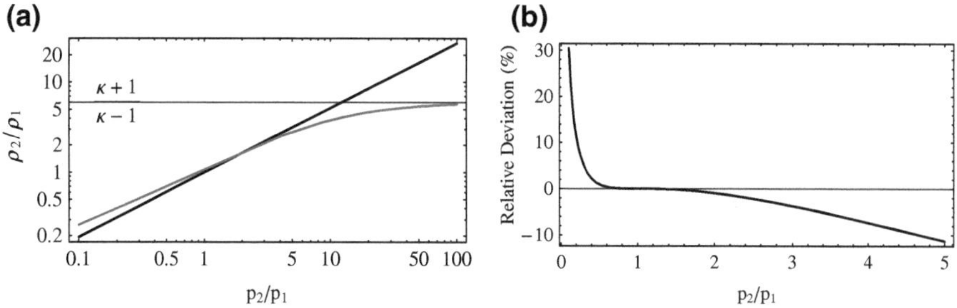

Fig. 2.1 Comparison between Rankine–Hugoniot and isentropic process for N_2 (perfect gas) with constant specific heat. **a** Rankine–Hugoniot (red) and isentropic curve (black). **b** Relative difference between Rankine–Hugoniot and isentropic curve

Rankine–Hugoniot curve is the correct one for compressions with large pressure ratios. For pressure ratios close to one the two curves coincide, Fig. 2.1b shows the relative difference between the two curves. Finally, for pressure ratios smaller than one, thus corresponding to expansion, all curves with a density-ratio larger than the one of the isentropic curve are prohibited by the second principal law because they would imply a decrease of entropy during the expansion. Since no assumption has been made so far concerning the distance between points (1) and (2), in principle the above results are applicable for large distances as well as for the limit of the distance going to zero. The latter one does actually occur in nature in the form of compression shocks in supersonic flows. Since the pressure jump in such a shock is usually comparable to the static pressure of the gas, it must always be considered a strong distortion. This implies that the Rankine–Hugoniot equations have to be used for the description of supersonic compression shocks.

Such shocks naturally occur always when a supersonic flow encounters some sort of obstacle in its path. Thinking first of a subsonic flow, it is clear that the flow will be influenced downstream *and* upstream therefore leading to a smooth adaptation of the flow that starts already well ahead of the position of the obstacle. This leads to a smooth transition from the unperturbed flow field far upstream to the deformed flow field close to the object. In the case of a supersonic flow, this is impossible because *smooth* upstream adaptation is equivalent to the upstream propagation of small perturbations that cause the gradual flow field deformation as the gas streams in from the unperturbed far field and gets closer to the obstacle. Only a *strong* perturbation in the form of a discontinuous shock is able to propagate with supersonic speed and, therefore, is able to propagate upstream in a supersonic flow. As it propagates its amplitude shrinks and so eventually it comes to a halt at a position where its propagation speed, given by (2.34), exactly matches the one of the supersonic flow. In this way a steady-state shock front is formed. A real-world example of such a shock front is visible in the simulation results presented in Sect. 3.2.6, Fig. 3.25 as well as in the experimentally obtained interferometric image displayed in Sect. 4.4, Fig. 4.5.

As explained above, expansion shocks never occur because they violate the second principle law of thermodynamics and the expansion takes place as a continuous isentropic state change.

So far, the shock front has been considered a steady-state phenomenon but it is, of course, valid to use a coordinate system that is moving with velocity w_1, thus, co-moving with the flow prior to the shock. Then the shock front appears to propagate into the undisturbed medium with the velocity $-w_1 = u$. For the perfect gas with constant specific heat, this propagation speed of the shock front can be expressed by

$$u = c\sqrt{1 + \frac{\kappa + 1}{2\kappa}\frac{p_2 - p_1}{p_1}} \tag{2.34}$$

As mentioned above, this formula allows one to estimate the position of a steady state shock that forms in front of an obstacle in the flow.

A more realistic case for a propagating shock would be the one of a tube that is split into two parts by a wall confining a perfect gas of a certain pressure on one side and a perfect gas with lower pressure on the other. The sudden removal of the membrane results in a shock that travels into the lower pressure medium. But now, since no additional gas is added on the high pressure side, the pressure drops there as the shock propagates. This case can be treated within the more general framework of (supersonic) wave propagation in perfect gases. It is described in textbooks, for example [1] and gives the following result in the case of vacuum on the low pressure side of the membrane

$$w = \frac{2}{\kappa - 1}c_0 \tag{2.35}$$

Here, c_0 is the sound velocity in the medium on the high pressure side prior to the removal of the wall and w is the velocity of the first disturbance that propagates into the vacuum. Given that the values of κ usually lie between 2.3 and 2.7, it is clear that this velocity can be significantly larger than the velocity of sound.

From the Eqs. 2.21–2.23 the following relations for the change of state in a shock front can be derived:

$$\frac{w_2}{w_1} = \frac{\rho_1}{\rho_2} = 1 - \frac{2}{\kappa + 1}\left(1 - \frac{1}{M^2}\right) \tag{2.36}$$

$$\frac{p_2}{p_1} = 1 + \frac{2\kappa}{\kappa + 1}(M^2 - 1) \tag{2.37}$$

$$\frac{T_2}{T_1} = \frac{c_2^2}{c_1^2} = \frac{1}{M^2}\left[1 + \frac{2\kappa}{\kappa + 1}(M^2 - 1)\right]\left[1 + \frac{\kappa - 1}{\kappa + 1}(M^2 - 1)\right] \tag{2.38}$$

$$M_2^2 = \frac{1 + \frac{\kappa - 1}{\kappa + 1}(M^2 - 1)}{1 + \frac{2\kappa}{\kappa + 1}(M^2 - 1)} \tag{2.39}$$

$$\frac{s_2 - s_1}{c_v} = ln\left[1 + \frac{2\kappa}{\kappa+1}(M^2 - 1)\right] + \kappa ln\left[1 - \frac{2}{\kappa+1}\left(1 - \frac{1}{M^2}\right)\right] \tag{2.40}$$

As can be seen, for the perfect gas with constant specific heat, the state change that occurs in a shock depends on the initial Mach number $M = M_1$ only. Here, M is defined as $M = w/c$ where w is the local velocity and c the local speed of sound according to Eq. 2.31. It must be kept in mind that these formulas are valid only for initial Mach numbers $M >= 1$ since for smaller Mach numbers isentropic expansion occurs. Figure 2.2 shows plots of the relations (2.36–2.40) for Nitrogen (solid black line), Argon (dashed black line), and Helium (solid red line). Since Argon and Helium are both monoatomic gases and, hence, have the same specific heat ratio of 5/3 and because only ratios are plotted the lines for these two gases coincide in all the plots. Again, it can be seen that the density ratio converges towards an asymptotic value while the pressure and temperature ratios grow without limits. Figure 2.2 shows the important fact that the flow is always subsonic after the shock with post-shock mach numbers decreasing as initial mach numbers increase. The entropy change normalized to the heat capacity $(s_2 - s_1)/c_v$, Fig. 2.2, can be used as a measure for the strength of the shock. To this end, however, it is also possible to define "local" reservoir conditions of a flow $w_0 = 0, \rho_0, p_0, \ldots$ as those conditions reached by bringing the flow isentropically to rest, thereby constituting the ideal conditions that could optimally be reached. Here, optimally means that in the case of isentropic compression, no pressure drop occurs and the initial reservoir conditions are obtained again. So, by definition, in an isentropic flow, the local reservoir conditions are constant throughout the flow. For the non-isentropic shock it follows from energy conservation that the local reservoir temperature is also constant but the local reservoir pressure and density drop with the amount of the drop being proportional to the strength of the shock.

$$\frac{\rho_{0,2}}{\rho_{0,1}} = \frac{p_{0,2}}{p_{0,1}} = \left[1 + \frac{2\kappa}{\kappa+1}(M^2 - 1)\right]^{-\frac{1}{\kappa-1}}\left[1 - \frac{2}{\kappa+1}\left(1 - \frac{1}{M^2}\right)\right]^{-\frac{1}{\kappa-1}} \tag{2.41}$$

This equation is plotted in Fig. 2.2 showing that for low Mach numbers a shock is a quite efficient and for higher Mach numbers a quite inefficient way of decelerating and compressing a flow. Similar to the Eqs. 2.36–2.41, expressions for oblique shock fronts can be derived. The tangential velocity components are preserved by the shock for the shock-normal velocity components as well as for density, pressure, temperature, and local reservoir conditions, Eqs. 2.36–2.41 are still applicable by simply substituting $M \sin\gamma$ for M where $M = M_1$ is the Mach number before the shock and γ is the angle between the initial flow velocity and the shock front. Similarly to perpendicular shock fronts, the second principal law prohibits oblique expansion shocks as well. There are two extreme points for the pressure rise in the oblique shock, namely a maximum for $\gamma = 90°$ corresponding to a normal shock perpendicular to the flow and a minimum for $\sin\gamma = 1/M = \sin\alpha$ with α being the Mach angle. The latter represents the weakest possible distortion of a supersonic flow, so values of $M \sin\gamma < 1$ are not meaningful.

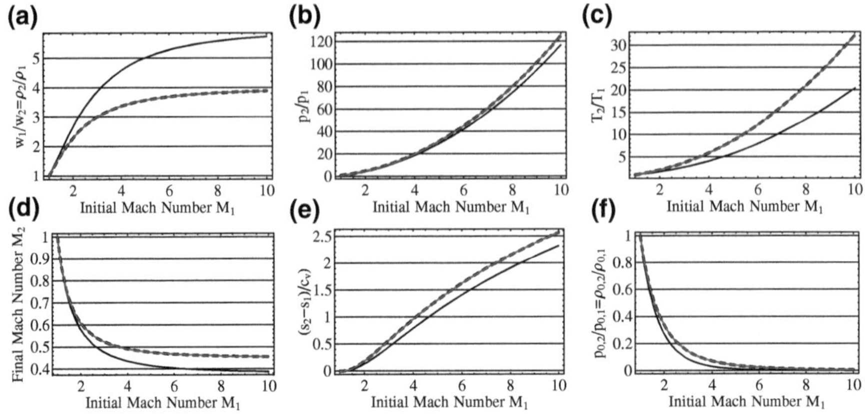

Fig. 2.2 Change of important state variables in a normal shock in a perfect gas (N_2,*solid black*, Ar, *dashed black*, He, *solid red line*) with constant heat capacity. **a** Density. **b** Pressure. **c** Temperature. **d** Mach Number. **e** Entropy. **f** Local Reservoir Pressure

2.1.4 Continuous Flows in Nozzles

In the following, only continuous flows within nozzles are considered. Of course this does not mean that discontinuous shocks cannot occur within nozzles. It will briefly be discussed below under which conditions this will happen. However, for the case of interest here, namely a nozzle that is attached to a vacuum chamber, it is clear that no shocks can occur because expansion shocks are impossible. Therefore, in the following only the continuous case is treated extensively.

In order to describe flows that are bound and guided by walls within a 1-D theory it is necessary to introduce the cross section f of the gas-duct into the governing equations. This is most easily done regarding the differential equations corresponding to Eqs. 2.21–2.23 [1]:

$$\frac{1}{w}\frac{\mathrm{d}w}{\mathrm{d}x}+\frac{1}{\rho}\frac{\mathrm{d}\rho}{\mathrm{d}x}+\frac{1}{f}\frac{\mathrm{d}f}{\mathrm{d}x}=0 \tag{2.42}$$

$$w\frac{\mathrm{d}w}{\mathrm{d}x}+\frac{1}{\rho}\frac{\mathrm{d}p}{\mathrm{d}x}=Y \tag{2.43}$$

$$w\frac{\mathrm{d}w}{\mathrm{d}x}+\frac{\mathrm{d}i}{\mathrm{d}x}=\frac{\mathrm{d}q}{\mathrm{d}x} \tag{2.44}$$

Here (2.42) is the continuity equation, now including the flow cross section f (m^2), (2.43) is the equation of motion including an external volume–force Y and (2.44) states energy conservation. In the case that there are no external forces and no energy or heat flows through the system boundaries, $Y=0, \mathrm{d}q/\mathrm{d}x=0$, this corresponds to a continuous, isentropic flow $ds/dx=0$. Then, by using (2.30) and $M=w/c$ Eq. 2.43 can be written as

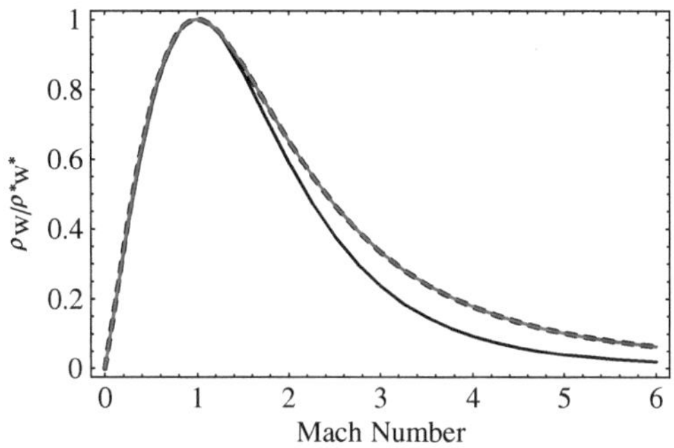

Fig. 2.3 Flow density normalized to critical values in a converging–diverging flow for N_2 *black solid*, Ar *black dashed* and He *red solid*

$$\frac{1}{\rho}\frac{\mathrm{d}\rho}{\mathrm{d}x} = -M^2\frac{1}{w}\frac{\mathrm{d}w}{\mathrm{d}x} \tag{2.45}$$

This shows that for small Mach numbers the relative variation of the density is smaller than the relative variation of the velocity and for large Mach numbers vice versa. This leads to the limits of uncompressible flow for very low Mach numbers, and, for high Mach nubmers, to hypersonic flows where the maximum velocity (2.24) has (almost) been reached and stays more or less constant and only density, pressure, and temperature vary strongly.

By combining now Eqs. 2.45 and (2.42) the following relationship between the Mach number and the flow cross section can be derived:

$$(1 - M^2)\frac{1}{w}\frac{\mathrm{d}w}{\mathrm{d}x} = \frac{1}{\rho w}\frac{\mathrm{d}(\rho w)}{\mathrm{d}x} = -\frac{1}{f}\frac{\mathrm{d}f}{\mathrm{d}x} \tag{2.46}$$

Here, ρw is a new parameter called the flow density and gives the total mass flow after multiplication by the flow cross section, $\dot{m} = \rho w f$. The total mass flow, of course, has to be constant throughout the flow if no sources or sinks are present. Inspection of Eq. 2.46 shows that for subsonic flows the velocity grows with shrinking cross section and that for supersonic flows it grows with growing cross section. This effect is exploited in converging–diverging de Laval nozzles as depicted in Fig. 2.3. First in the converging section the flow accelerates up to $M = 1$ which is reached in the throat. Then, in the diverging nozzle section, the flow is allowed to expand further thereby acquiring supersonic speeds corresponding to $M > 1$. For $M = 1$ the flow cross section evidently has a minimum and the flow density a maximum and, therefore, this point in the flow is of special importance because it separates the sub- from the supersonic regime. The flow parameters that the gas obtains at that point are called "critical" parameters and are signed with an asterisk. These critical values can be calculated for the perfect gas with constant heat capacity as follows: Starting from Eq. 2.23, using (2.13) and assuming that the gas is initially at rest $w_0 = 0$ one gets

$$\frac{w^2}{2} + c_p T = c_p T_0 \tag{2.47}$$

Here, variables with subscript 0 denote initial (reservoir) values. With (2.25) and (2.31) this transforms to

$$w^2 + \frac{2}{\kappa - 1}c^2 = w_{max}^2 \tag{2.48}$$

Here, the known critical values for M and w, namely $M^* = 1$, $w^* = c^*$ can be introduced leading to

$$w_{max}^2 = \frac{\kappa + 1}{\kappa - 1}(w^*)^2 = \frac{\kappa + 1}{\kappa - 1}(c^*)^2 = \frac{2}{\kappa - 1}c_0^2 = (\kappa + 1)c_p T^* = 2c_p T_0 \tag{2.49}$$

and, by exploiting the usual isentropic Eqs. 2.18–2.20, finally the following equations are obtained that now relate the relevant critical flow parameters to the reservoir values:

$$w^* = \sqrt{\frac{\kappa - 1}{\kappa + 1}2c_p T_0} = c_0\sqrt{\frac{2}{\kappa + 1}} \tag{2.50}$$

$$\frac{p^*}{p_0} = \left(\frac{2}{\kappa + 1}\right)^{\frac{\kappa}{\kappa - 1}} \tag{2.51}$$

$$\frac{\rho^*}{\rho_0} = \left(\frac{2}{\kappa + 1}\right)^{\frac{1}{\kappa - 1}} \tag{2.52}$$

$$\frac{T^*}{T_0} = \frac{2}{\kappa + 1} \tag{2.53}$$

$$\frac{\rho^* w^*}{\rho_0 c_0} = \left(\frac{2}{\kappa + 1}\right)^{\frac{\kappa + 1}{2(\kappa - 1)}} \tag{2.54}$$

Now that the critical parameters are available, Eqs. 2.13, 2.18 and 2.22 can be used to calculate the flow density in the whole flow domain. Then, one arrives at the following result:

$$\frac{f^*}{f} = \frac{\rho w}{\rho^* w^*} = M\left[1 + \frac{\kappa - 1}{\kappa + 1}(M^2 - 1)\right]^{-\frac{\kappa + 1}{2(\kappa - 1)}} \tag{2.55}$$

A plot of this function is given in Fig. 2.3. Due to the continuity relation $w\rho f = w^*\rho^* f^*$ the vertical axis in Fig. 2.3 can also be read as the cross section ratio f^*/f thereby showing that for each cross section two solutions are obtained, one corresponding to the subsonic and the other to the supersonic case. From $M = 0$ to $M = 1$ the flow accelerates towards the critical cross section and its flow density grows to its maximum value. As has been shown above, this requires a decreasing flow cross section that finally reaches a minimum value—the nozzle throat—where critical values are obtained. From there on, in order to further increase the Mach number, the flow cross section has to increase again in order to allow for the additional expansion and the corresponding reduction of the flow density that is necessary to reach the supersonic regime. This converging–diverging nozzle is called a de Laval nozzle and is a very frequently used device for producing supersonic gas flows. The mass flow through the nozzle is now given by $\dot{m}^* = \rho^* w^* f^*$ which evaluates to

$$\dot{m}^* = f^* \sqrt{\kappa \left(\frac{2}{\kappa+1}\right)^{\frac{\kappa+1}{\kappa-1}} p_0 \rho_0} \tag{2.56}$$

while the mass flow in general is expressed by

$$\dot{m} = f \sqrt{\frac{2\kappa}{\kappa-1} p_0 \rho_0} \left(\frac{p}{p_0}\right)^{\frac{1}{\kappa}} \sqrt{1 - \left(\frac{p}{p_0}\right)^{\frac{\kappa-1}{\kappa}}} \tag{2.57}$$

Since $\dot{m}^* = \dot{m}$ holds everywhere in the flow domain, relation (2.57) implicitly defines the pressure p as a function of the cross section f in the whole flow. Comparable to Fig. 2.3, for the Mach Numbers—this equation has two solutions for p for each cross section f, one on the supersonic and one on the subsonic side of the critical cross section in the nozzle throat. With Eqs. 2.19, 2.20 the density ρ and the temperature T are calculated and the flow velocity w can be easily obtained via mass conservation from $w = \dot{m}/(\rho f)$.

Figure 2.4, shows plots of the most important flow parameters along a de Laval nozzle that is connected to a reservoir containing air at a pressure of 50 bar, a temperature of 300 K (corresponding to an initial density of 1.2×10^{21} cm^{-3}), and no initial velocity. The nozzle contour is also shown. It consists of a converging part with straight conical walls that is 1.75 mm long and has an entrance diameter of 3 mm. The throat diameter is 1 mm and subsequently the nozzle expands again with a straight conical contour to an exit diameter of 3 mm.

Tables 2.1 and 2.2 display numeric values of important flow parameters in the reservoir, at the throat where critical values are obtained and at the nozzle exit of a de Laval nozzle with a cross section ratio f_{exit}/f^* of 9. The pressure drops by almost a factor of 2 from the reservoir to the nozzle throat and subsequently in the supersonic section by a factor of 62. The density drop is less pronounced and reaches almost exactly a factor of 30 at the exit of the nozzle. The difference between the density and the pressure drop is explained by the fact that also the temperature decreases strongly—to 77 K at the nozzle exit—so that for the density that is a function of pressure and temperature some part of the pressure drop is compensated by the temperature drop. It is also interesting to note that the exit velocity of 670 m/s is already 86% of the theoretical maximum velocity (2.25) of 777 m/s. The mass flow through the nozzle amounts to 9.16 g/s which corresponds to a volume flux of 0.16 l/s at reservoir conditions (50 bar, 300 K). To give an impression of the influence of the type of gas used as a medium, in Table 2.2, the flow parameters for Helium are given using the same reservoir values. The medium enters the equations through two parameters, the molecular weight M and

Fig. 2.4 Variation of flow parameters inside a de Laval nozzle with a cross section ratio $f_{exit}/f^* = 9$. Reservoir parameters: $p_0 = 5 \times 10^6$ Pa = 50 bar, $T_0 = 300$K, $w_0 = 0$, medium: Helium. The parameters plotted are: nozzle radius *nozzle contour*, pressure *black*, and density *red* normalized to their respective reservoir values, temperature (K), Mach number (1) and flow velocity (m/s)

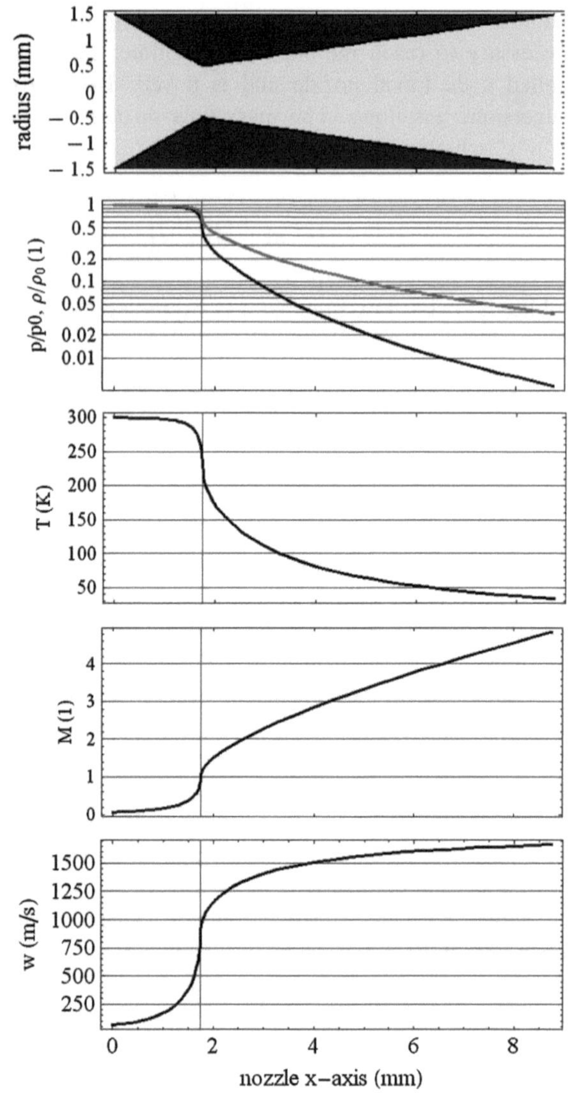

the heat capacity (for constant pressure) c_p. For air, the average molecular weight is 28.9696 g/mol and c_p is1006.43 J/kg K, helium has a molecular weight of 4.0026 g/mol and a c_p of 5193 J/kg K. Inspection of the equations show, however, that in almost all the cases the parameter κ—which is a function of both c_p and M—is the only medium-dependent parameter. As mentioned above, the significance of κ lies in the fact that $2/(\kappa - 1)$ corresponds to the number of degrees of freedom of the gas molecules. Its value is 2.4 for Air and 2.67 for helium which corresponds to 5 degrees of freedom for the *average* air molecule and 3 degrees of freedom—only the translations—for the point-like helium atom. The main

Table 2.1 Flow variables in a de Laval nozzle

Parameter	Reservoir	Throat	Nozzle exit
Pressure (bar)	50	26.4	0.429
Density (10^{19} cm^{-3})	121	76.5	4.02
Temperature (K)	300	250	77.2
Mach Number (1)	0	1	3.80
Velocity (m/s)	0	317	670

Medium: air. Entrance diameter 3 mm, throat diameter 1 mm. Exit diameter 3 mm. Reservoir parameters: $p_0 = 5 \times 10^6$ Pa = 50 bar, $T_0 = 300$ K, $w_0 = 0$ m/s

Table 2.2 Flow variables in a de Laval nozzle

Parameter	Reservoir	Throat	Nozzle exit
Pressure (bar)	50	24.4	0.218
Density (10^{19}cm^{-3})	121	78.4	4.627
Temperature (K)	300	225	34.1
Mach Number (1)	0	1	4.84
Velocity (m/s)	0	882.6	1662

Medium: helium. Entrance diameter 3 mm, throat diameter 1 mm. Exit diameter 3 mm. Reservoir parameters: $p_0 = 5 \times 10^6$ Pa = 50 bar, $T_0 = 300\,\mathrm{K}$, $w_0 = 0\,\mathrm{m/s}$

differences between the air flow and the helium flow are the following: the smaller molecular weight of He manifests itself in a much higher exit velocity of 1,662 m/s which compares to a theoretical maximum velocity of 1,765 m/s. The smaller heat capacity leads to a much smaller exit temperature of 34.1 K as compared to 77.2 K for air which also leads to a higher Mach number for helium of 4.84. The density at the exit is almost the same, the pressures differ by a factor of 2. The mass flow is smaller by a factor of 2.5 for Helium but the volume flow is higher by a factor of 2.8—another effect of the small molecular weight of helium. If the jet is assumed to emanate into vacuum it is not possible within this simple model to predict its evolution outside the nozzle because the flow cross section is not a priori known and cannot easily be calculated. Also, within the nozzle the presented model is accurate only as long as the part of the flow that is affected by the presence of walls is negligibly small in comparison to the flow dimensions. Since the processes within the wall-affected zone, the so-called boundary layer, depend non-trivially on flow parameters like pressure, pressure gradient, velocity, fluid viscosity, and turbulence, there are no sufficiently accurate analytical models that would allow to study this phenomenon in analytical fashion. Therefore, one has to rely on numeric simulation of the flow which will be the central point of the following chapter.

To conclude this section, Fig. 2.5, gives an overview over the dependence of important state variables at the exit of the de Laval nozzle on backing pressure, cross section ratio between nozzle throat and exit, reservoir temperature, and on the gas type. In all the plots, the plot for Helium is the solid red, the one for Argon the dashed black, and the one for N_2 the solid black line. Since Argon and Helium are both monoatomic gases with the same isentropic exponent it is expected that

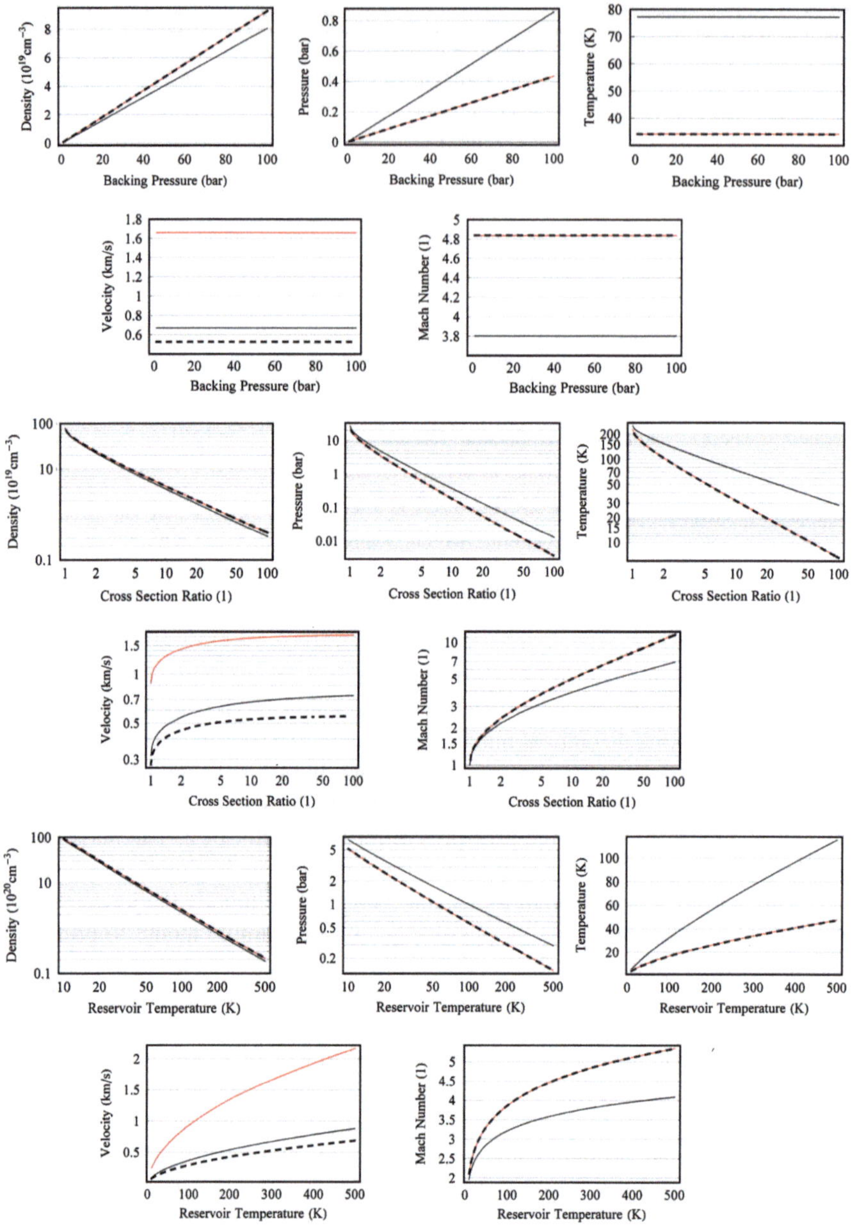

Fig. 2.5 Variation of relevant state variables with backing pressure *upper five plots*, ratio between exit and throat cross sections *middle five plots*, and reservoir temperature *lower five plots*. Medium: Helium *solid red*, Argon *dashed black*, N_2 *black*

they show similar behavior. For the upper five and the lower five plots, a cross section ratio of nine was used, for the middle five plots the backing pressure was set to 50 bar. Looking first at the upper five plots in the figure it can be seen that the (particle) density and pressure vary linearly with backing pressure. Since density is measured in particles per cm^3 and not in kg/m^3, the two lines for Argon and Helium overlap showing exactly the same result. The (static) pressure at the nozzle exit also depends linearly on the backing pressure and, again, Argon and Helium give the same results. Temperature, velocity, and Mach number do not depend on the backing pressure at all. It is apparent that the exit velocity is the only parameter that actually depends on the molecular weight of the gas showing different results for Argon and Helium.

Turning now to the middle five plots, it is clear that by changing the cross section ratio between the nozzle exit and throat, the flow parameters at the nozzle exit readily vary over orders of magnitude, therefore, all plots are in double logarithmic scale. Again, the exit velocity is the only parameter where Argon and Helium show different results. The almost linear behavior of density, pressure, and temperature at the nozzle exit imply a power law dependence on the cross section ratio. The exit velocity rapidly converges to the maximum attainable speed which is given by energy conservation (see 2.25). The fact that the Mach number keeps growing is explained by the falling temperature that lowers the speed of sound.

The lower five plots show the dependence of the flow conditions at the nozzle exit on the temperature of the gas reservoir. Particle density and pressure follow almost perfectly a power law (double logarithmic plots). Interestingly, the exit temperature varies rather gently with the initial temperature. Since it can be seen that velocity and, therefore, also the Mach number vary with the reservoir temperature, it represents a convenient way of tuning the velocity and Mach number at the nozzle exit. Especially the latter one can be of importance because in the case that supersonic shock fronts are used in the experiment, the ratio of all the flow parameters before and after the shock only depend on the Mach number (see Eqs. 2.36–2.40).

2.1.5 Cluster Formation in Supersonic Gas Jets

It is a long known fact that high pressure gas jets emanating into vacuum can be a formidable way of producing large clusters of atoms or molecules [2–6]. This is due to the high densities and low temperatures that are simultaneously reached in such jets. Since de Laval nozzles are especially well suited for reaching both—high density and low pressure—at the nozzle exit, they are especially effective tools for cluster formation. Since the presence of clusters in a gas jet may have an influence on any experiments conducted with these jets, a short analysis of the cluster production in supersonic gas jets is presented.

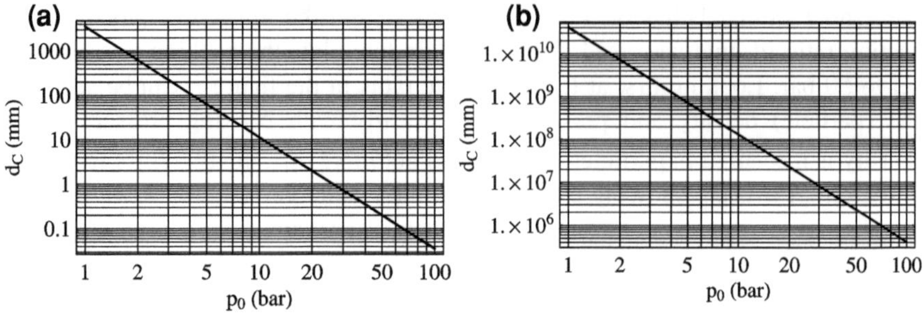

Fig. 2.6 Plots of critical nozzle diameters and backing pressures that lead to a dimer mole fraction of 0.01 which is considered as the onset of clustering. **a** Argon. **b** Helium

Since dimers are the first step in cluster formation, the dimer mole fraction X_{A2} represents a meaningful measure for the number of clusters to be expected in a gas jet. The dimer formation goes on continuously from the nozzle throat diameter downstream until the so-called sudden freeze surface is reached beyond which no more significant condensation into clusters occurs any more [3]. The position of this surface is essentially a function of temperature and can be assumed to lie several nozzle diameters away from the nozzle exit in the case of small diameter ratios (<4:1) of the de Laval nozzle. Therefore, the final dimer mole fraction can be regarded an upper limit of the dimer content close to the nozzle exit. In order to calculate the final dimer mole fraction the following empirical formula is given in [3]:

$$X_{A2} = \frac{1}{2}\left(\rho_P \sigma^3 \left(\frac{\varepsilon}{k_B T}\right)^{\frac{7}{5}} \left(\frac{d^*}{\sigma}\right)^{\frac{2}{5}}\right)^{\frac{5}{3}} \tag{2.58}$$

Here, ρ_{P} is the gas number density, σ is the atom size, ε is the potential well depth, k_{B} is Boltzman's constant, d^* is the critical (smallest) diameter of the nozzle. Values for these parameters can be found in [3] and [7]. For Helium the following values are used: $\sigma = 2.66 \times 10^{-10}$m, $\varepsilon/k_B = 10.9\,$K, and for Argon $\sigma = 3.33 \times 10^{-10}$m, $\varepsilon/k_B = 144.4\,$K. In the case of a de Laval nozzle for d^* the following expression is substituted [3], [4]:

$$d^*_{eq} = 0.736\, d_C \cot(\alpha), \quad \gamma = 5/3 \tag{2.59}$$

$$d^*_{eq} = 0.866\, d_C \cot(\alpha), \quad \gamma = 7/3 \tag{2.60}$$

where d_{eq^*} is an equivalent diameter that depends on the gas type defined by the specific heat ratio γ, the half opening angle α, and the throat diameter d_{C}. According to [2, 3, 8], considerable clustering sets in for mole fractions larger than 0.01. This is shown in Fig. 2.6, where those value pairs of d_{C} and p_0 are plotted that correspond to a value of X_{A2} of 0.01. In order to have no clusters, parameters below the lines have to be chosen. As can be seen, this is posing some limits on pressure and/or nozzle diameter for Argon whereas for Helium no clusters can be expected in a realistic pressure and diameter range. For de Laval nozzles, the expressions given by Eqs. 2.59 and 2.60 must be substituted for d_C.

References

1. Oswatitsch, K.: Gasdynamik. Springer, Heidelberg (1952)
2. Knuth, E.L.: Size correlations for condensation clusters produced in free-jet expansions. J. Chem. Phys. **107**(21), 9125–9132 (1997)
3. Knuth, E.: Dimer-formation rate coefficients from measurements of terminal dimer concentrations in free-jet expansions. J. Chem. Phys. **66**(8), 3515–3525 (1977)
4. Hagena, O.F.: Nucleation and growth of clusters in expanding nozzle flows. Surf. Sci. **106**, 101 (1981)
5. Bauer, S.H., Chiu, N-S., Wilcox, C.F. Jr.: Kinetics of condensation in supersonic expansion (ar). J. Chem. Phys. **85**(4), 2029–2037 (1986)
6. Hillenkamp, M., Keinan, S., Even, U.: Condensation limited cooling in supersonic expansions. J. Chem. Phys. **118**(19), 8699–8705 (2003)
7. Hirschfelder, J.O., Curtiss, C.F., Bird, R.B.: Wiley, (1964)
8. Scheier, P., Märk, T.D.: Isotope enrichment in ne clusters. J. Chem. Phys., **87**(9), 5238–5241 (1987)

Chapter 3
Numeric Flow Simulation

3.1 Flow Models for Computational Fluid Dynamics

The analytical model describing the flow of a newtonian fluid consists of a set of three coupled differential equations: the continuity relation together with the Navier–Stokes equations and the energy conservation.

$$\frac{\mathrm{D}\rho}{\mathrm{D}t} = -\rho\,\mathrm{div}\vec{w} \tag{3.1}$$

$$\rho\frac{\mathrm{D}\vec{w}}{\mathrm{D}t} = \vec{f} - \mathrm{grad}p + \mathrm{Div}\left[\mu\left(2\dot{\varepsilon} - \frac{2}{3}\delta\mathrm{div}\vec{w}\right)\right] \tag{3.2}$$

$$\rho c_p\frac{\mathrm{DT}}{\mathrm{D}t} = \mathrm{div}(\lambda\,\mathrm{grad}T) + \beta T\frac{\mathrm{Dp}}{\mathrm{D}t} + \Phi \tag{3.3}$$

Here, the substantial derivative (D/Dt) is defined as

$$\frac{\mathrm{Da}}{\mathrm{D}t} = \left(\frac{\partial a}{\partial t} + \vec{w}\cdot\mathrm{grad}a\right),\quad \frac{\mathrm{D}\vec{b}}{\mathrm{D}t} = \left(\frac{\partial\vec{b}}{\partial t} + \mathrm{grad}\left(\frac{1}{2}\vec{w}^2\right) - \vec{w}\times\ \mathrm{curl}\vec{b}\right) \tag{3.4}$$

$\vec{f}$ is an external volume-force like for example gravity, μ is the viscosity in the newtonian relation for the frictional tensions and the tension tensor τ and the deformation rate tensor $\dot{\varepsilon}$ are given by

$$\tau = \mu\left(2\dot{\varepsilon} - \frac{2}{3}\delta\mathrm{div}\vec{w}\right) \tag{3.5}$$

$$\dot{\varepsilon} = \begin{pmatrix} \frac{\partial w_1}{\partial x_1} & \frac{1}{2}\left(\frac{\partial w_2}{\partial x_1} + \frac{\partial w_1}{\partial x_2}\right) & \frac{1}{2}\left(\frac{\partial w_3}{\partial x_1} + \frac{\partial w_1}{\partial x_3}\right) \\ \frac{1}{2}\left(\frac{\partial w_1}{\partial x_2} + \frac{\partial w_2}{\partial x_1}\right) & \frac{\partial w_2}{\partial x_2} & \frac{1}{2}\left(\frac{\partial w_3}{\partial x_2} + \frac{\partial w_2}{\partial x_3}\right) \\ \frac{1}{2}\left(\frac{\partial w_1}{\partial x_3} + \frac{\partial w_3}{\partial x_1}\right) & \frac{1}{2}\left(\frac{\partial w_2}{\partial x_3} + \frac{\partial w_3}{\partial x_2}\right) & \frac{\partial w_3}{\partial x_3} \end{pmatrix} \tag{3.6}$$

K. Schmid, *Laser Wakefield Electron Acceleration*, Springer Theses,
DOI: 10.1007/978-3-642-19950-9_3,

Here, $\dot{\boldsymbol{\varepsilon}}$ is given in cartesian coordinates. $\boldsymbol{\delta}$ is the Kronecker tensor defined as $\delta_{ij} = 1$ for $i = j$, and $\delta_{ij} = 0$ for $i \neq j$. The dissipation function Φ is given by

$$\Phi = \operatorname{div}(\boldsymbol{\tau}\vec{w}) - \vec{w}\,\mathrm{Div}\boldsymbol{\tau} \tag{3.7}$$

and the thermal expansion coefficient β is defined as

$$\beta = \frac{1}{\rho}\left(\frac{\partial\rho}{\partial T}\right) \tag{3.8}$$

λ ($[\lambda]$ = J/ms K) is the heat conductivity coefficient in the heat transport equation

$$\vec{q} = -\lambda\,\mathrm{grad}T \tag{3.9}$$

The derivation of the above formulae can be found, for example, in [1, 2]. Here, only the area of their applicability is shortly discussed and a brief introduction to their numerical solution is given. The Eqs. 3.1, 3.2, and 3.3 are a system of five-coupled non-linear differential equations for p, T and the three components of $\vec{w}$. To close the equations, the equation of state $\rho(p, T)$ is needed and the parameters defining the properties of the medium $c_p(p, T)$, $\mu(p, T)$ and $\lambda(p, T)$. The fluid is assumed to be a continuum. This means that the smallest relevant scales of the flow—usually these are the smallest-scale turbulences—are much larger than the molecular mean free path. The fluid is newtonian, isotropic and its tension-tensor is symmetric. The hypothesis of Stokes holds (no relaxation processes) and time derivatives do not influence the equation of motion (principle of locality). The local thermodynamic state of the medium can be described by an equation of state of two variables, for instance, p and T. Sources of heat, heat-radiation, and diffusion processes are not included in this model. The complexity of the system of Eqs. 3.1–3.3 makes it very difficult to find analytic solutions. So far, these have only been found for strongly idealized and very simple cases. Especially, the appearance of turbulence—which unfortunately is the typical case in engineering applications—is very hard to tackle analytically. This leaves numeric solution as the only viable way to obtain results concerning real flows. Also, here the appearance of turbulence effects causes this task to be quite difficult because, in order to be accurate, the numeric model has to capture the macroscopic scales of the flow *and* the microscopic scales of the smallest scale turbulences. This causes the necessary resolution of the computational grid to be prohibitively large in many cases. A short overview over possible solutions to this problem are given in Sect. A.0.2.

3.1.1 Parameterization of de Laval Nozzles

If not otherwise stated, the de Laval nozzles analyzed in the following sections have the generic shape shown in Fig. 3.1. For Details on the nozzle design see Appendix B.

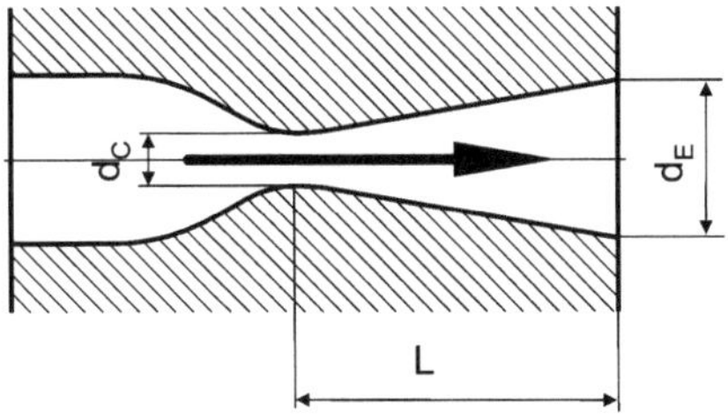

Fig. 3.1 Parameterization of de Laval Nozzles. If not otherwise stated, the de Laval nozzles studied in the following have this generic shape consisting of a divergent section with conical walls that is connected by a radius of four times the throat diameter d_C to the nozzle throat. The radius on the high pressure (*left*) side of the nozzle throat is two times d_C. In the parameter studies presented in Sects. 3.2.3 and 3.2.4, the nozzles are parameterized by the throat diameter d_C, the exit diameter d_E, and the length of the conical section L

3.1.2 Size Effects and Effects of Low Pressure

The continuum approximation which is the foundation of all the models incorporated in Fluent requires that the smallest geometric feature or flow feature is much larger than the molecular mean free path. A dimensionless parameter that measures the ratio between a geometric dimension and the mean free path is the Knudsen number Kn, which is defined by

$$Kn = \frac{\lambda_{MF}}{L} = \frac{1}{\sqrt{2}\sigma\rho_P L} \tag{3.10}$$

Here, λ_{MF} is the molecular mean free path, σ is the scattering cross section of the molecules or atoms of the fluid, ρ_P is the number density, and L is a geometric length. The right-hand side in the equation is only true for perfect gases. σ has been calculated from $\sigma = r_W^2\pi/4$, thus using r_W for defining a hard-sphere potential, where r_W is the Van der Waals radius of Helium which is 140 pm. Using the Knudsen number, four different flow regimes can be distinguished, they are listed in Table 3.1. The term "slip boundary" in Table 3.1 refers to a technique which allows to extend the range of applicability of the Navier Stokes equations up to Knudsen numbers of 0.1. In the slip regime, the gas–phase velocity at a solid surface differs from the velocity at which the wall moves. Furthermore, the gas temperature at the surface differs from the wall temperature. This velocity slip and temperature jump can be modelled and included into the Navier Stokes solver. It should be mentioned, that for the transition regime there exist the Burnett [3] equations which allow again to augment the Navier Stokes equations and, therefore, use continuum solvers in the transition regime which still is computationally very expensive for direct simulation, especially in 3D [4, 5]. Assuming that the 1D isentropic theory of gas flow is a good approximation at the nozzle center, Eq. 3.10

Table 3.1 Flow regimes according to Knudsen number

Kn	Flow regime	Numerical approach
$Kn \to 0$	Continuum	Euler equations
$< 10^{-3}$	Continuum	Navier Stokes equations
$10^{-3} \leq Kn < 10^{-1}$	Slip flow	Navier Stokes equations with slip boundary
$10^{-1} \leq Kn < 10$	Transition	Direct simulation Monte Carlo methods
$Kn > 10$	Molecular	Direct simulation Monte Carlo methods

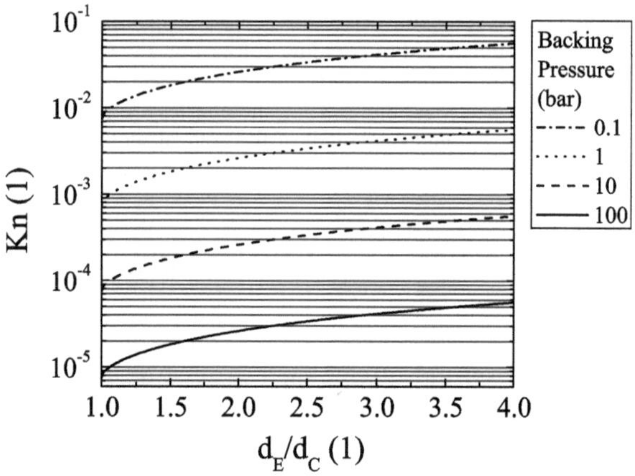

Fig. 3.2 Local Knudsen number plotted along the diverging section of the de Laval nozzle for different backing pressures. Even for pressures as low as 1 bar it is justified to use the continuum model

allows to calculate the Knudsen number without any simulations for different backing pressures.

Figure 3.2 shows results for a nozzle with $d_C = 1$ mm, $d_E = 3$ mm and L = 6 mm for backing pressures of 0.1, 1, 10 and 100 bar. It can be seen that only for backing pressures lower than 10 bar, Kn gets larger than 10^{-3} close to the exit of the nozzle. This is just at the border of the slip flow regime, but it has been verified that the difference between the solutions of the Navier Stokes equations with and without boundary slip condition for de Laval nozzles are negligible at least up to Knudsen Numbers of 0.002. This also goes conform with results obtained in [6, 7]. Furthermore, because Kn depends linearly on the characteristic length L (here the throat diameter), Fig. 3.2 also can be used to determine how small a nozzle can still be modelled correctly within the continuum approximation. The figure shows that down to a backing pressure of 1 bar, a nozzle with a throat diameter of 1 mm, an exit diameter of 3 mm, and a length of the diverging section of 6 mm is within the continuum regime. Therefore, a nozzle ten times smaller will be in this regime down to a pressure of 10 bar. Also the smallest nozzle simulated here, with a throat diameter of 0.05 mm, can safely be modelled using the continuum approximation down to a backing pressure of 50 bar. It should be noted that, of course, the expansion taking place inside the de Laval nozzle is different for different expansion ratios. However, as can be seen in Fig. 3.2, the Knudsen number does not change too much along the diverging section of the de Laval nozzle on the

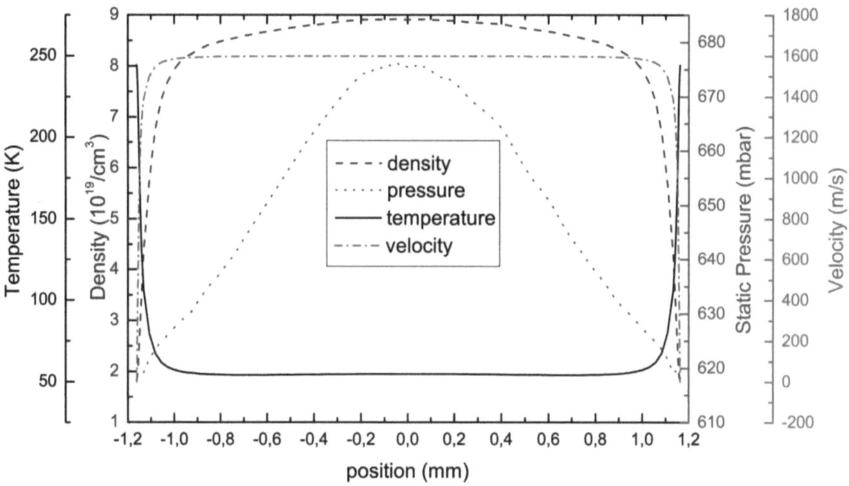

Fig. 3.3 Profiles of density, pressure, temperature inside the diverging section of a de Laval nozzle 2 mm before the nozzle exit

logarithmic scale, and, therefore, the same minimum pressures that have been identified above for the nozzle with expansion ratio 9 should be also valid for the other simulated expansion ratios of 4 and 16.

3.1.3 Boundary Layers

The major issue for microscopic nozzles are boundary layers that form along the internal walls of the nozzle. The boundary layer that forms here is a thin flow layer attached to the nozzle walls that connects the high velocity flow in the volume of the domain with the flow boundary that is immediately attached to the confining walls, and therefore has zero velocity. Thus, the most intuitive way of defining the boundary layer thickness is to define it as the region close to the wall where the velocity changes from 0 to 90% of the value in the center of the nozzle. A typical velocity profile 2 mm before the exit of the nozzle is shown in Fig. 3.3. This is the typical flat-top velocity profile for turbulent flows with very thin boundary layers. However, also other flow parameters can be used to define a boundary layer, namely the temperature and the density. Also, for those parameters, profiles are shown in Fig. 5.6. In fact, it is the density-boundary layer which is more useful for the consideration of free jets because there is a smooth transition from density-boundary layer thickness to the width of the density gradient of the free jet. It is important to note that the core flow and the boundary layer are two physically very distinct domains. The first one can be quite accurately described by the Euler equations which neglect completely the viscosity of the flow while in the latter, viscosity and friction play a dominant role and are actually the only reasons for the formation of the boundary layer at all. For example, it can be seen in Fig. 3.3 that the temperature rises by almost 200 K within the boundary layer. This is pure frictional heat generated inside the flow, the walls in the model are adiabatic. The boundary layer has essentially two important effects on the gas flow in nozzles: first, it is the boundary layer thickness which governs the steepness of the density gradient of the gas jet very close to the nozzle exit. Second, it can be seen in Fig. A.4 how the boundary layer thickness increases along the expanding section of the de Laval nozzle which is mainly due to the density drop and the increasing velocity caused by the expansion of the flow within the conical part of the nozzle. Therefore, it is clear that in order to obtain very steep density gradients, one must aim for low velocities (or low Mach numbers) and high pressure. For a de Laval nozzle this leads to small expansion ratios which is in conflict with the fact that for large expansion ratios the jet has a smaller divergence angle. Since the boundary layer builds up along the wall also short nozzles are favorable for thin boundary layers at the nozzle exit. At a fixed expansion ratio, this leads to larger cone angles of the de Laval nozzle and can also lead to transverse pressure modulations, see Sect. 3.2. Therefore, there is a tradeoff between gradient steepness and collimation/quality of the jet. Quality here refers to the flatness of the flat-top density profile at the exit of the nozzle. The second effect of the boundary layer is that it blocks some fraction of the cross section of the nozzle. Depending on backing pressure and nozzle size, this can lead to a slightly modified "effective" nozzle geometry with little or no influence on the overall flow properties or to "chocking" of the flow. The latter means that the boundary layer starts to influence strongly all flow parameters when there is at least one point in the flow domain where the distance between two confining walls becomes comparable to the boundary layer thickness. To quantify this, the displacement thickness δ_D can be used. It is defined as

$$\delta_D = \int_0^{\delta_{99}} \left(1 - \frac{u(y)\rho(y)}{u_\infty \rho_\infty}\right) dy \tag{3.11}$$

and measures by how much the flow streamlines are pushed away from the walls by the viscous boundary layer. δ_{99} is the (normal) distance from the wall where the velocity has 99% of its value far away from the wall and u_∞ and ρ_∞ are the velocity and the density far away from the wall.

3.2 Simulation Results

3.2.1 Supersonic Flows and the de Laval Nozzle

Supersonic gas jets generated by de Laval nozzles are ideal targets for laser–plasma experiments due to their flat-top density profiles and steep density gradients at the jet edges, both of which are impossible to produce with a (subsonic)

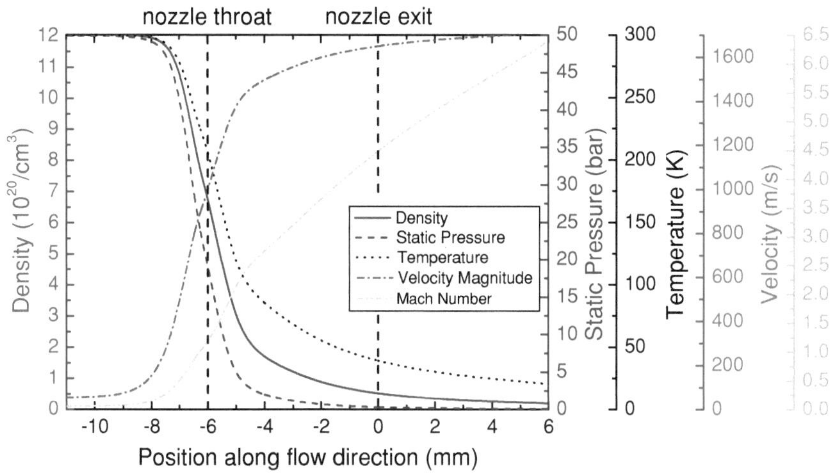

Fig. 3.4 Density, pressure, temperature, velocity magnitude and Mach number plotted along the axis of a de Laval nozzle with a throat diameter of 1 mm, an exit diameter of 3 mm and a length of the diverging nozzle section of 6 mm. Medium: Helium, backing pressure: 50 bar

cylindrical nozzle. The variation of the most important flow parameters along the nozzle axis of a de Laval nozzle with a throat diameter of 1 mm, an exit diameter of 3 mm and a length of the diverging nozzle section of 6 mm are shown in Fig. 3.4. The medium is Helium with a backing pressure of 50 bar. For comparison with a cylindrical nozzle, approximate values for all flow parameters shown in Fig. 3.4 for the case of a cylindrical nozzle can be obtained by simply taking the values at the nozzle throat. This is a quite good approximation because approximately there the flow becomes supersonic and becomes, therefore, independent of upstream conditions. Since for the de Laval nozzle the expansion of the gas mainly happens inside the guiding nozzle wall, the acceleration caused by the expansion is strongly guided into forward direction. The lower temperature at the nozzle exit of the de Laval nozzle, in the given example in Fig. 3.4 40 K as compared to 200 K, leads to the fact that at the same density the pressure is much lower at the exit of the de Laval nozzle. Since the pressure induces the transversal spread of the gas jet it is, therefore, clear, that the supersonic jet emanating from a de Laval nozzle will diverge much less than the one from a cylindrical nozzle at the same density. This also leads to the fact that the flat-top density profile at the exit of the conical de Laval nozzles studied here is preserved over a significant propagation distance. Figure 3.5 shows a comparison between a cylindrical subsonic nozzle with a diameter of 0.75 mm and a de Laval nozzle with a throat diameter of 0.25 mm and an exit diameter of 0.75 mm. For both nozzles, the gas is Helium with 50 bar backing pressure. Only right at the nozzle exit is the subsonic nozzle able to maintain a steep gradient but even there the profile is not really flat-top. Only 0.1 mm away from the nozzle exit, for the subsonic nozzle, the gas already

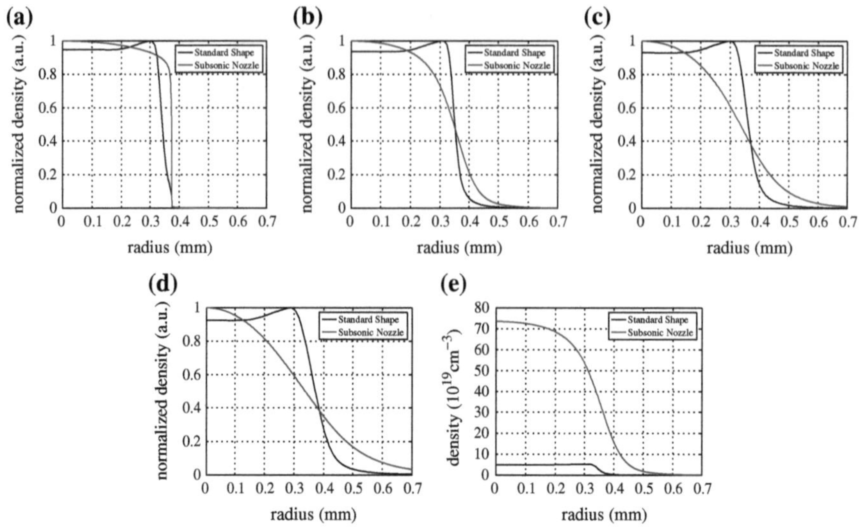

Fig. 3.5 Comparison of density line-outs at distances of 0, 0.1, 0.2, 0.3 mm from the nozzle exit. *Black line-outs* correspond to a de Laval nozzle with 0.25-mm throat and 0.75-mm exit diameter, *red line-outs* to a cylindrical subsonic nozzle with a diameter of 0.75 mm. Plots **a–d** are normalized to one to demonstrate the qualitative differences between the profiles, plot **e** is not normalized and shows that the cylindrical nozzle produces far higher densities than the de Laval nozzle. The backing pressure in both cases was 50 bar (Helium)

expanded into all directions producing broad gradients and a convex central shape. And 0.2 mm from the nozzle exit—this is 27% of the diameter—the density line-out produced by the subsonic nozzle is already a very good approximation to a gaussian. By contrast, the supersonic jet produced by the de Laval nozzle maintains its flat-top profile over a distance of more than its exit diameter of 0.75 mm. Figure 3.5e shows that this comes at a price, however.

The expansion of the gas that takes place inside the de Laval nozzle lowers the density at the nozzle exit significantly-in this case, it is approximately an order of magnitude lower than the density produced by the subsonic nozzle. It has been stated that for the subsonic nozzle the gas expands transversally much stronger than for the supersonic nozzle. This can be seen in Fig. 3.6 where the free jet divergence of the two nozzles is compared. The reason for the much larger divergence of the jet from the subsonic nozzle is the high pressure and comparatively low velocity at the nozzle exit. The gas leaves the nozzle and accelerates transversally, rapidly approaching an almost isotropic velocity distribution corresponding to a gaussian density profile. Only during the short period that the gas needs to accelerate transversally to speeds similar to the longitudinal one, there is a significant departure from this behavior.

For the de Laval nozzle, things are quite different. By the time the gas leaves the nozzle, it has already converted a substantial part of its enthalpy into kinetic energy and the confining nozzle walls guided this expansion into the forward

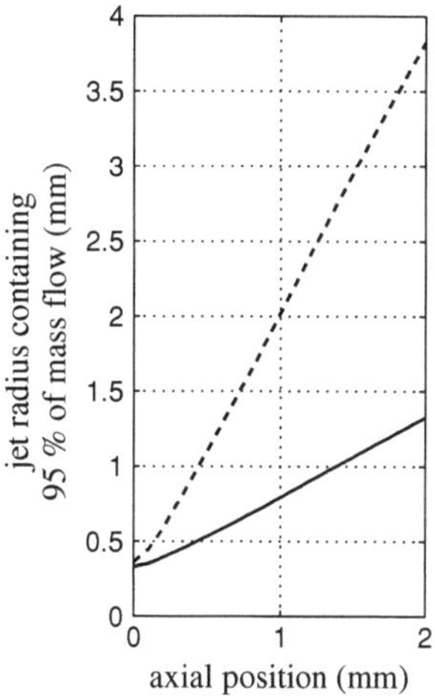

Fig. 3.6 Divergence of jets emanating from a supersonic de Laval nozzle (*solid line*) and from a subsonic nozzle (*dashed line*). The gas jet from the subsonic nozzle is much less collimated, having a full opening angle of the 95% mass flow contour of 122°. This compares to 56° for the de Laval nozzle

direction. Therefore, the exit velocity is much larger (approximately a factor of 2 in this case) and the pressure is much smaller (approximately a factor of 20) so that the gas has no chance any more to accelerate transversally to velocities similar to its forward velocity. An approximate value for the divergence of the supersonic jet can be obtained by calculating the transversal spread velocity by Eq. 2.35 using the sound speed at the nozzle exit. For Helium and a de Laval nozzle that has an exit diameter of three times its throat diameter and a reservoir temperature of 300 K this gives a full divergence angle of 63° which is actually quite close to the values obtained by simulation. Also far away from the nozzle exit, this divergence is approximately preserved because due to the low pressure at the exit of the nozzle within a very short distance from the nozzle the additional expansion that occurs is sufficient to render the jet essentially collisionless and the particles follow ballistic trajectories, therefore, preserving the collimated velocity distribution generated by the de Laval nozzle.

Here, the nomenclature should be clarified: the isentropic expansion that takes place after the gas has left the cylindrical "subsonic" nozzle rapidly accelerates the flow to supersonic speed. Therefore, by speaking of a subsonic nozzle or a subsonic gas jet or a supersonic nozzle, one always refers to the maximum Mach number that the gas acquires *inside* the nozzle itself. Outside of it—provided that the gas emanates into vacuum, or at least a sufficiently low pressure surrounding medium—supersonic conditions are always obtained.

The simplicity of the geometry of the cylindrical nozzle makes it quite easily possible—in contrast to the jet from a de Laval nozzle—to obtain fitting formulas that allow to calculate all important flow parameters outside the nozzle, Miller in [8, 9]. It should be noted, however, that the calculations presented there neglect the existence of boundary layers and, therefore, can be applied only to cases with high backing pressure and/or large nozzle diameters.

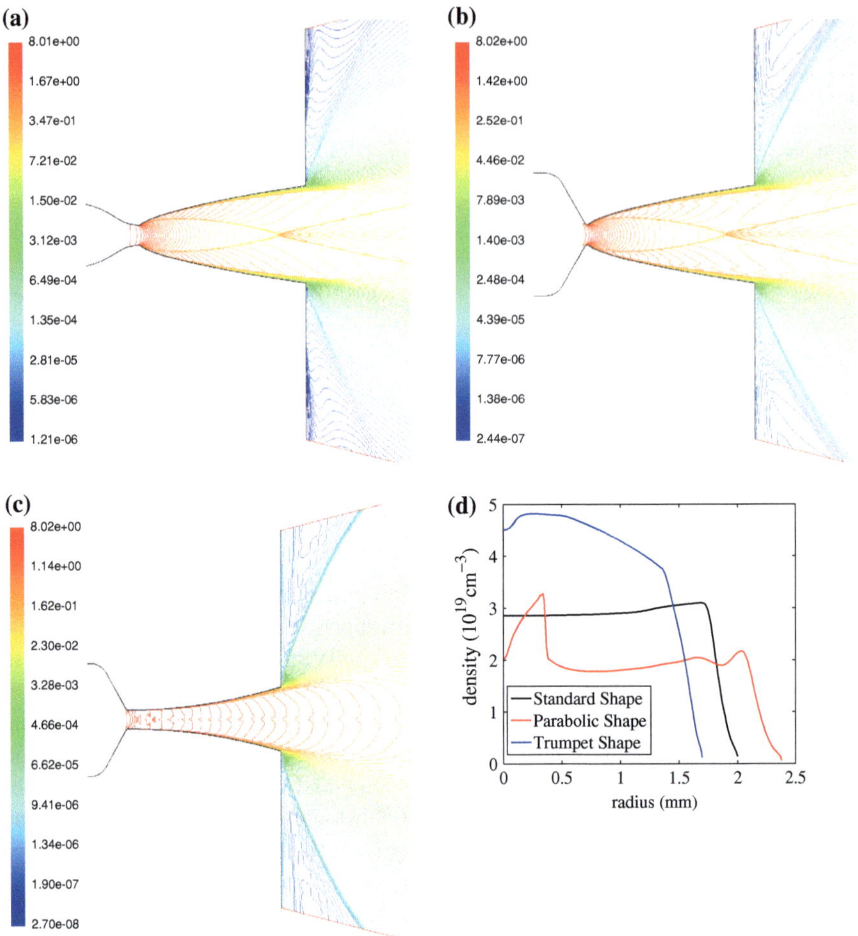

Fig. 3.7 Density contours on logarithmic scale for two nozzle with parabolic shape (**a**), (**b**) and one trumpet shaped nozzle (**c**). Only the convergent part of the two parabolic nozzles are different, the supersonic sections of the nozzles are identical. Evidently, the flow properties do not depend on the geometry of the subsonic part of the nozzle. Inlet pressure is in all cases 50 bar, throat diameter 1 mm, exit diameter 2.4 mm for the parabola nozzles and 1.7 mm for the trumpet nozzle. **d** Line-outs of density at a distance of 1 mm from the nozzle exit for the standard conical nozzle shape (*black*) the parabolic nozzle (*red*) and the trumpet-shaped nozzle (*blue*). Rather strong flow distortions for the parabolic nozzle and a non-flattop profile for the trumpet shaped nozzle rule these designs out for laser–plasma experiments

3.2.2 Optimal Nozzle Shape

Besides the standard conical shape also trumpet-like nozzles, Fig. 3.7c, and parabolic nozzles, Fig. 3.7a and 3.7b were considered. The parabolic nozzle contour was generated by setting the angle of the wall at each position within the supersonic section of the nozzle to the local Mach-angle at this position. This gives an approximate parabola-shape. The two parabolic nozzles shown differ from each other only in the geometry of the convergent part of the nozzle in order to demonstrate that the supersonic flow properties depend only weakly on subsonic upstream parameters. The emergence of strong shocks immediately after the throat is evident, indicating that even though the Mach number is only slightly above 1 immediately after the throat, still the gas cannot follow the rather abrupt cross-section change. By contrast, the trumpet shaped nozzle shows a very smooth flow in the supersonic section but does not produce the flat-top profile desirable for laser–plasma experiments. Line-outs of density at a distance of 1 mm from the nozzle exit for the different nozzle shapes are depicted in Fig. 3.7d. The parabolic nozzle shape leads to rather strong density jumps (red plot) caused by the shock fronts produced inside the nozzle, the trumpet shaped nozzle produces smooth but non-flat-top profiles (blue plot).

Therefore, in the following only the standard conical nozzle shape has been taken into account and studied extensively because apparently only this geometry produces the desired flat-top density profiles. In order to establish an optimum nozzle shape, a parameter study has been conducted, varying the expansion ratio E, the length of the diverging section of the Nozzle L, the size of the whole nozzle, and the shape of the nozzle contour. Regarding the optimal general nozzle shape, the simulations show that it is preferable to have smooth round contours rather than sharp edges, at least at the throat and downstream of it, because in this region the flow is sonic or supersonic and, therefore, forms shock fronts at sharp edges. Therefore, in order to prevent the development of strong shock fronts or expansion fans after the nozzle throat, the throat-contour of all the nozzles was rounded with a radius of four times the throat diameter on the supersonic side and two times it on the subsonic side.

3.2.3 Influence of the Nozzle Geometry on the Flow Parameters

The parameter study described in the following uses de Laval nozzles as shown in Fig. 3.1 characterized by the length L of the diverging nozzle section, the throat diameter d_C, and the exit diameter d_E. The set of nozzles considered in this study is summarized in Table 3.2.

Here, α is the full opening angle of the diverging section of the nozzle as it results from L, d_C and d_E. In all the cases, the medium was Helium with a backing pressure of 50 bar and a background pressure of 10^{-2} mbar. The k-ω shear stress transport turbulence model was used with the coupled implicit steady state 2D-axisymmetric solver. Grid sizes ranged from 370,000 to 730,000 cells. For further processing after the simulation concluded in FLUENT, line-outs of temperature, pressure, density, Mach number, axial and radial velocity where computed along lines parallel to the r-axis in 0.1 mm steps along the x-axis. For the computation of

Table 3.2 Parameter study: nozzle geometry—list of simulations

Nr. (1)	d_C (mm)	d_E (mm)	L (mm)	α (°)
1	1	2	2	14.0
2	1	2	3	9.46
3	1	2	4	7.13
4	1	2	6	4.76
5	1	3	4	14.0
6	1	3	6	9.46
7	1	3	8	7.13
8	1	3	12	4.76
9	1	4	6	14.0
10	1	4	9	9.46
11	1	4	12	7.13
12	1	4	18	4.76

the displacement thickness also wall-normal line-outs of those parameters were produced with the same spacing. This data was exported to ascii files and then imported into MATLAB for further processing. Data analysis was done mainly concerning three parameters: first, the divergence of the free jet that emanates from the nozzle because flow-directionality is an important experimental parameter. Second, the flow displacement inside the nozzle because it determines the influence of the walls onto the core flow, and third, the density gradient inside and outside of the nozzle because it is of primary interest for the laser–plasma experiments that the gas jets studied here are used for.

The results for the free jet divergence are discussed first: in the hypothetic case that the gas flow expands so strongly inside the nozzle that it obtains its maximum velocity (1.24), it would just continue outside the nozzle with the jet boundaries being a prolongation of the nozzle walls. In reality this is not quite the case but still at the nozzle exit the gas flow has suffered a pressure drop of almost two orders of magnitude with respect to reservoir values and a temperature drop to values around 50 K. Most of the enthalpy of the gas has been converted to kinetic energy and the confining nozzle walls have guided the flow mainly into forward direction. The rest of the enthalpy at the nozzle exit is now converted to kinetic energy outside of the nozzle. Without the guiding walls, the velocity increases not only in longitudinal but also in transversal direction, thereby leading to a continuing expansion of the jet that goes on until density and temperature drop to the point where no collisions between the particles occur any more. From there on, each particle then follows a ballistic trajectory. Out of these basic considerations, it is clear that collimated jets can be expected from nozzles with large internal expansion of the flow, which is accomplished by large ratios between the areas of the critical cross section and the exit cross section (expansion ratio). On the other hand, it can be expected that the opening angle of the diverging part of the nozzle will also influence the divergence of the free jet. Figure 3.8 shows the results of the parameter study about the two parameters diameter ratio d_C/d_E and opening angle

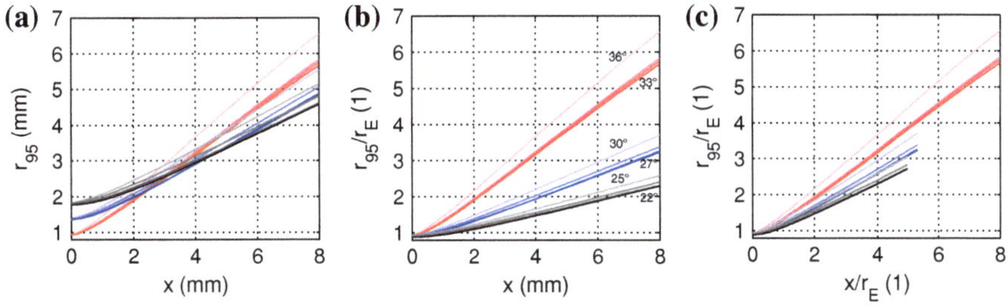

Fig. 3.8 Free jet radius containing 95% of the mass flow for different ratios of the critical cross section to the exit cross section. **a** Free Jet Radius, **b** Free Jet with radial axis normalized to exit radius, **c** Free Jet with both axes normalized to exit radius. *Grey shades* are for an expansion ratio of 1:16, *blue shades* for 1:9 and *red shades* for 1:4. Four cases are given for each cross section ratio referring to different full opening angles (or length) of the diverging section of the de Laval nozzle, from darker to lighter shades 9.6°, 14.4°, 19.3°, 29.3°. Divergence half-angles of the jets are given in (**b**)

α of the de Laval nozzle. The different diameter ratios correspond to the shades of red (1:2), blue (1:3), and grey (1:4) in Fig. 3.8. The opening angles of 9.6°, 14.4°, 19.3°, and 29.3° are represented by the saturation of the color in the figure, with the largest opening angle corresponding to the lightest shade and the smallest opening angle to the darkest shade. The plot depicts the radius of the gas jet that contains 95% of the mass flow versus distance from the nozzle exit. In Fig. 3.8a the expected result that the gas jet is more collimated for larger expansion ratios can be seen. Since the critical cross section was kept constant with a diameter of 1 mm, the exit diameter of the nozzle varied from 2 (1:2) to 3 (1:3) and 4 (1:4) mm. It can be seen that after approximately 3 mm propagation the jet produced by the nozzle with the largest exit diameter has a smaller diameter than the jets of the other two nozzles. The picture gets clearer if one normalizes the radial axis to the respective exit radius of each nozzle, which is shown in Fig. 3.8b. It is obvious that the nozzle with the largest expansion ratio (black) produces the jet that increases its radius with the slowest rate. By normalizing both axes to the exit diameter (Fig. 3.8c) it can be seen that this effect does not simply scale with the exit diameter, also in this normalized plot the largest expansion ratio produces the most collimated jet. Concerning the opening angle of the de Laval nozzle, it can be seen that its influence is actually quite small for values below approximately 20° in all the cases. The full divergence angles for all the cases are given by the values in Fig. 3.8b, they range from 45° for the 1:4 nozzles to up to 72° for the 1:2 nozzles and are calculated for the part of the free jet between 5 and 7 mm distance from the nozzle exit. This compares to values of 51°, 61° and 78.5° calculated with formula (2.35) for the 1:2, 1:3 and 1:4 nozzle, respectively. The differences are mainly explained by the fact that the case described by formula (2.35) oversimplifies the problem and, furthermore, it is valid in the one dimensional case, not in the cylindrical geometry of this problem. Still, the very simple model shows at least qualitative agreement and provides some justification to the

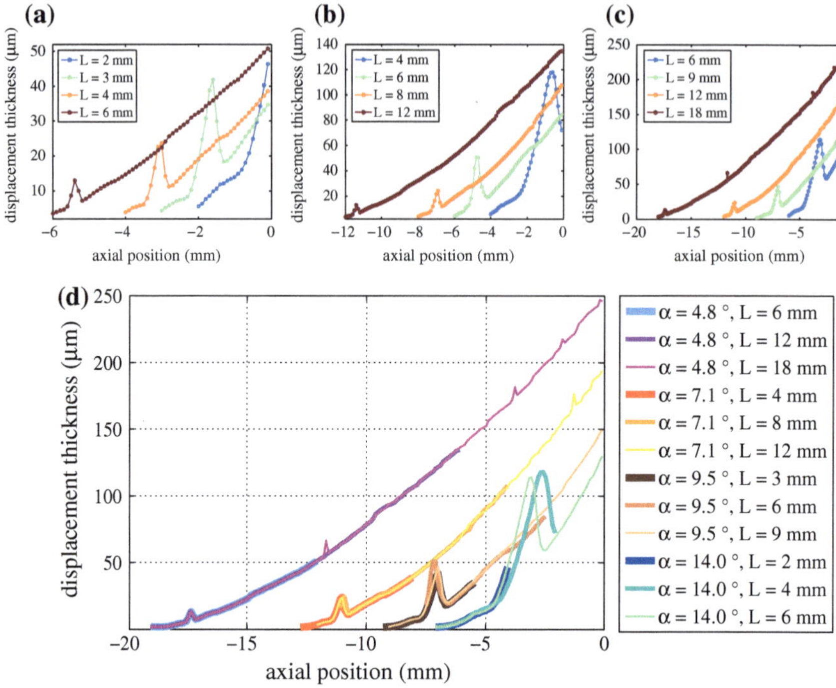

Fig. 3.9 Plots **a–c** show the displacement thickness as it develops inside the diverging section of the nozzle for a ratio between the throat and the exit diameter of 1:2, 1:3 and 1:4, respectively. The displacement thickness grows both with the length of the nozzle and the ratio. Plot **d** shows the same data arranged by grouping together the same angles rather than the same ratios

explanation of the process of jet expansion: the transversal expansion of the jet into the surrounding vacuum takes place at supersonic speed and therefore leads to more divergent jets than one would expect if the jet only expanded transversally at the speed of sound.

Another interesting parameter to be known as a function of nozzle geometry is the displacement thickness as defined by formula (3.11) in Sect. 3.1.3. As detailed there, it is the amount by which the flow is displaced from the nozzle walls by the boundary layer that builds up in the gas flow close to the wall. Reducing the size of nozzles is useful only down to the point where the displacement thickness at some point in the nozzle becomes comparable or even as large as the local nozzle radius. At this point, the mass flow through the nozzle is strongly reduced and all flow properties are strongly altered, preventing the development of the desired supersonic flow with the typical flat-top density profile. Panel 3.9 shows plots of the displacement thickness versus axial position. Figures 3.9a–c show these plots grouped together for nozzles with equal ratio d_C/d_E. The nozzle exit always lies at x = 0 in these plots. It is obvious that the displacement thickness grows with both the length L of the diverging section of the nozzle and the diameter ratio. Since the set of nozzles in this study (Table 3.2) is constructed such that for each of the four angles there are three nozzles, one with 2, one with 3 and one with 4-mm exit diameter, and all having the same throat diameter of 1 mm, it is clear that actually the nozzle contours are congruent. This is to say that nozzle Nr. 8 can be constructed by cutting nozzle Nr. 12 after 12 mm of its length. In the same way nozzle 4 can be constructed by cutting Nr. 8 or Nr. 12 after 6 mm. Since the flow in the diverging section of the nozzle is supersonic it cannot depend on downstream conditions and therefore it should be expected that the flow in nozzle Nr. 4 looks exactly like the flow in the first 6 mm in nozzles Nr. 8 and 12. Similarly, the flow in the first 3 mm of nozzles Nr. 6 and 10 should be equal to the flow in nozzle Nr. 2 and so on for all the four angles. This is shown in Fig. 3.9d where all the nozzles with ratios of 1:2 and 1:3 have been shifted along x such that their nozzle contours fit exactly their counterpart with ratio 1:4. The described correspondence between the cases with the same angle but different length is very nicely reproduced small deviations appearing only for the case $\alpha = 14°$. The peak that appears in all the cases approximately 1 mm after the nozzle throat is probably caused by the large velocity gradients there. The size of the peak gets amplified by the expansion that took place in the nozzle up to that point which is why it appears to be growing with nozzle angle. The three additional small peaks in the $\alpha = 4.8°$, L = 18 mm case and the one additional peak in the $\alpha = 7.1°$, L = 12 mm case are artifacts produced by data processing, they are not physical.

The following conclusions can be drawn from these results: the displacement thickness grows with propagation length along the wall and with the expansion that the gas undergoes during the propagation. In order to keep the displacement thickness small, short, and weakly expanding nozzles are preferable. Since the pressure modulation that always occurs shortly after the nozzle throat has an increasing impact on the size of the flow displacement with increasing nozzle angle (or decreasing length at the same diameter ratio) the nozzle should be kept long enough and the angle small enough to limit this effect and confine it well inside the nozzle. The plots for $\alpha = 14°$ interestingly show some differences between each other right at the position of that peak, so evidently small changes in initial parameters here are amplified strongly and lead to clearly distinguishable flow displacements. Since this could, therefore, be a possible source of instability, this provides another reason for keeping the nozzle long and/or the nozzle angle small enough.

For laser–plasma experiments, one of the most important parameters of a gas target is the width of the region within which the density rises from vacuum to the nominal value. In the following, this density gradient width is defined differently inside and outside the nozzle: inside it, the gradient width is taken as the distance from the wall where the density reaches 90% of its maximum value at each given x-position.

Since the density line-outs along lines of constant x show partly considerable deviations from a flat-top inside the nozzle, the results for the density gradient were much smoother and more realistic when actually only the region of the flow that is sufficiently close to the wall was taken into account for computing the maximum density. This region was defined as twice the distance from the wall where the velocity reaches 95% of its maximum value. Since the velocity profiles are flat-top along the whole nozzle length, this provides a reliable measure for the size of the wall-influenced region. Outside the nozzle, the gradient width is defined as the width of density rise from 10% to 90% of its maximum value. Panel 3.10 shows the results for the gradient width along the diverging section of the de Laval nozzle. The step-like features that are visible in almost all of the plots are artifacts and not physical, they are also visible in Fig. A.4a. Figure 3.10a is arranged similar to Fig. 3.9d and shows the same trends, but it seems that the density gradient is less sensitive to the fluctuations close to the nozzle throat. Since the displacement thickness depends on density and velocity only [formula (3.11)] this is a clear indication that the peaks that show up in the displacement thickness are actually caused by the velocity gradient close to the nozzle throat. Figure 3.10a shows a very smooth behavior of the density gradient growth inside the nozzle and the plots actually also are almost parallel to each other. Only nozzle Nr. 9 seems to slightly deviate from this behavior. The reason for this is not entirely clear but one possible explanation is that, in this case, the onset of a turbulent boundary layer can be seen. Hence, it is in principle a physical effect that is captured in the simulation. However, inspection of Fig. A.4a reveals that the k-ϵ-RNG and the k-ϵ-Realizable turbulence models predict a sudden rise in the displacement thickness, that could be explained by the onset of a turbulent boundary layer, already for a nozzle corresponding to nozzle Nr. 7, with a nozzle angle of only 7.13°. This is an expression of the fact that in general the prediction of the exact location of the transition from laminar to turbulent boundary layer is quite difficult and therefore this question can only be answered conclusively by measurement.

Figure 3.10b again shows the same plots as Fig. 3.10a but the plots are now shifted back to their original positions such that the nozzle exit is at x = 0 mm. Clearly, there is a trend that the density gradient grows with L and with d_C/d_E. Finally, Fig. 3.10c shows that the density gradients of the free jet follows a similar trend as its radius (Fig. 3.8). Also here, the density gradient width grows for larger expansion ratios and shows little dependence on the nozzle angle. The inset magnifies the first millimeter after the nozzle exit and reveals that for the nozzles with diameter ratios of 1:3 and 1:4, the density gradient width is actually constant for the first few hundred micrometers after the nozzle exit. This, of course, is a property highly appreciated in experiments because it makes the target properties independent of the distance from the nozzle within this range. Coming back now to Fig. 3.10b, it can be seen that the gradient width at the nozzle exit—so the points where the plot lines touch the 0 mm axis—follows, a quite regular pattern. This, of course, with the exception of nozzle 9 that has been detailed above. And indeed, if one plots the density gradient thickness at the nozzle exit versus the parameter $L \times d_E$ an almost linear dependence is revealed. This is shown in Fig. 3.11. The off-lying point at $L \times d_E = 20\,\mathrm{mm}^2$ corresponds to nozzle 9 and should probably not be taken into account. It is remarkable how the points for two completely different nozzles fall together at $L \times d_E = 11\,\mathrm{mm}^2$.

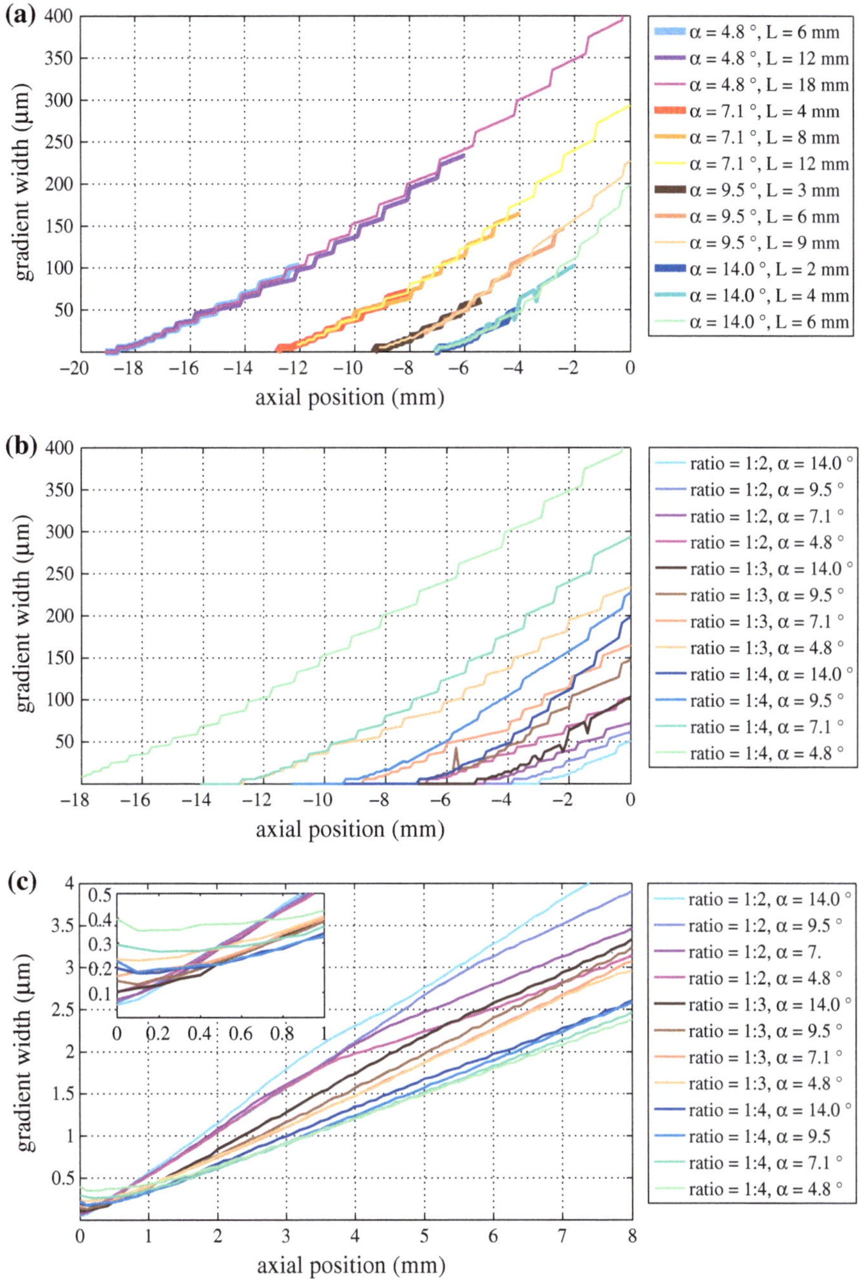

Fig. 3.10 Plot **a** shows the gradient width as it grows along the diverging section of the nozzle. The plots are grouped together by nozzle angle. Plot **b** shows the same data but grouped by diameter ratio. Plot **d** shows the development of the density gradient for the free jet. **a** Density gradients inside nozzle, grouped by nozzle angle, **b** density gradients inside nozzle, grouped by diameter ratio, **c** density gradients of the free jet, grouped by diameter ratio

Fig. 3.11 Width of the density gradient at the nozzle exit versus the parameter $L \times d_C/d_E$. A linear fit to the data is given as the red line

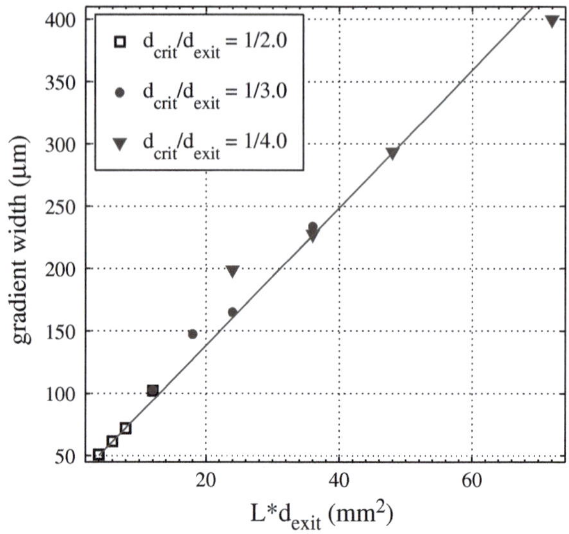

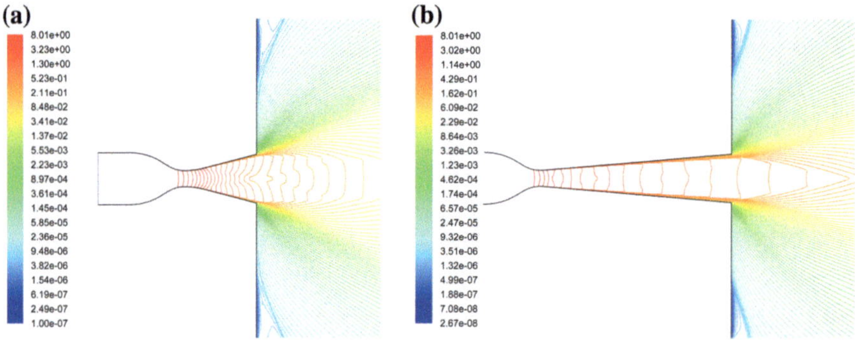

Fig. 3.12 Comparison of density contours inside two nozzles with different angles but the same diameters. The nozzle with the larger angles generates flow distortions close to the throat that propagate through the entire nozzle. For the smaller angle this is largely suppressed. **a** Nozzle with throat 1 mm, exit diameter 3 mm and nozzle angle of 14°, **b** Nozzle with throat 1 mm, exit diameter 3 mm and nozzle angle of 4.7°

In order to quantify the quality of the desired flat-top density profile, an analysis of the deviation from an ideal flat-top of the density profiles produced be the different nozzles was conducted. To illustrate the necessity of such an analysis, Fig. 3.12 shows constant density contours for two nozzles with different angles. It is obvious that the larger angle causes flow distortions that propagate upstream and lead to deviations from the flat-top profile at the nozzle exit. Figure 3.13 gives the results for all the studied nozzle shapes, with Fig. 3.13a showing the normalized density profiles at the nozzle and Fig. 3.13b showing the relative peak-to-peak fluctuation of density line-outs of the core flow taken along lines parallel to the radial axis. Here "core flow" is defined as the flow with a wall distance of at least

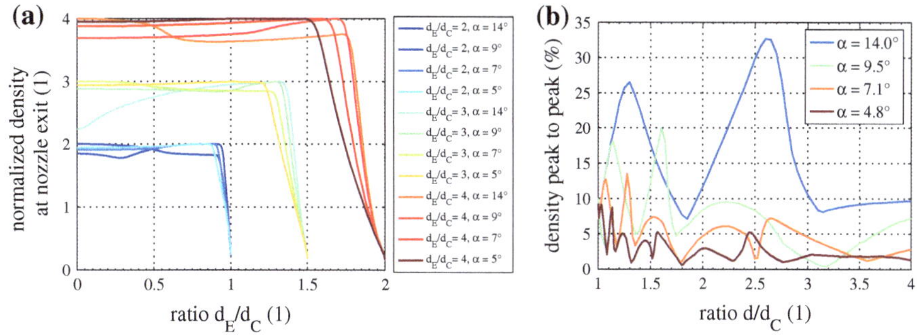

Fig. 3.13 Comparison of density line-outs along lines of constant x at the nozzle exit (**a**) and relative density peak-to-peak variation along the diverging section of the nozzles (**b**). The x-axis in (**b**) is given as the ratio between the local nozzle diameter at a given x-position and the throat diameter d_C. This implicitly defines the x-position but normalizes it in a way that all plots have the same length on the x-axis. It can be seen that for *(full)* nozzle angles of more than approximately 10° rather strong flow distortions can occur. **a** Normalized density line-outs at nozzle exit, **b** relative density peak-to-peak fluctuation of the core flow

two times the density gradient width. To improve clarity, the density line-outs in Fig. 3.13a are normalized to different values. Quite obviously, for (full) nozzle angles of more than 10° rather strong flow distortions occur. Figure 3.13b demonstrates that the distortions that have their origin close to the nozzle throat (Fig. 3.12), decay as the flow propagates along the diverging section of the de Laval nozzle with the decay rate depending inversely on the nozzle angle. So, in principle, for good flat-top quality one would opt for rather long nozzles with small angles, but of course this comes with the tradeoff of ever increasing boundary layer thickness, that grows linearly with the nozzle length, Fig. 3.11.

3.2.4 Effects of Nozzle Size and Varying Backing Pressure

A parameter study concerning the effects of changes in reservoir pressure and nozzle size on the flow properties is presented. It makes sense to combine these two parameters into one study because—as explained in Sect. 3.1.2—changing the size of a nozzle and changing the backing pressure both influences the flow behavior mainly through the varying influence of the presence of the walls. Per unit time a particle undergoes a certain number of collisions with other particles and a certain number of collisions with the walls. The ratio between these two collision rates determines the degree to which a flow is influenced by the presence of the walls. In this sense, walls that are close to each other or low gas pressure that leads to long mean free path length have the same effect: the particles collide more frequently with the walls and less frequently with each other, thereby enlarging wall effects. The dimensionless quantity that is a measure of the ratio of the

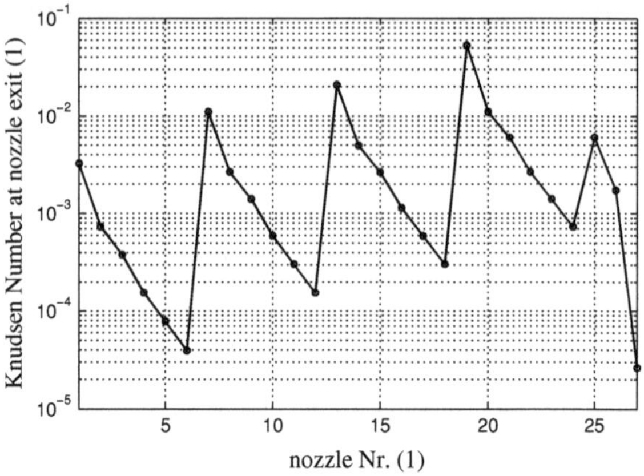

Fig. 3.14 Kn of nozzles in the pressure/size study

particle—wall and particle—particle collision rates is the Knudsen number Kn (Eq. 3.10).

The set of nozzles listed in Table 3.3 was considered in this study all of them being scaled versions of a nozzle with $d_C = 1\,\text{mm}$, $d_E = 3\,\text{mm}$ and L = 6 mm corresponding to a nozzle angle of 9.46°, see Fig. 3.1. As can be seen in Fig. 3.14, the Knudsen number Kn varies over almost four orders of magnitude within this set of simulations. The values for Kn shown in the figure and in Table 3.3 are calculated at the nozzle exit where Kn reaches its maximum value within the nozzle (see Fig. 3.2). Therefore, it allows not only to follow the variation of the flow properties with varying Knudsen number, but also gives an indication into which flow regime according to Table 3.1 the given case falls. As in the section above, the divergence of the free jet, as defined by the radius containing 95% of the mass flow, is considered. The results are shown in Fig. 3.15 ordered with respect to the parameter $p_0 \times d_C$. This parameter grows with shrinking Knudsen number at the nozzle exit, however, in contrast to Kn, the parameter $p_0 \times d_C$ is readily available for any nozzle and, thus, more useful when it comes to design decisions.

As can be seen in Fig. 3.15a, the jet is confined in a cone around the axis in all the cases. Smaller nozzle size or lower backing pressure lead to increasing jet divergence. The inset in Fig. 3.15a shows a magnification of the region close to the nozzle exit. Already inside the nozzle, the density gradient between the wall and the main flow becomes larger and larger for smaller pressures and smaller sizes leading to smaller jet radii at the nozzle exit. This indicates a strong reduction of the open cross sections within the nozzles caused by the development of ever thicker boundary layers. Figure 3.15b plots the full divergence angles calculated from the data presented in Fig. 3.15a between $x = 1.6 \times d_E$ and $x = 2.3 \times d_E$. For values of $d_C \times p_0$ between 100 and 10 (mmbar), the divergence angle is more or less constant around 56° and then rises up to a maximum value of 85° for $d_C \times p_0 = 0.13\,\text{mmbar}$. For even smaller pressures-size products, the flow becomes dominated by its boundary layers and the behavior changes significantly. In order

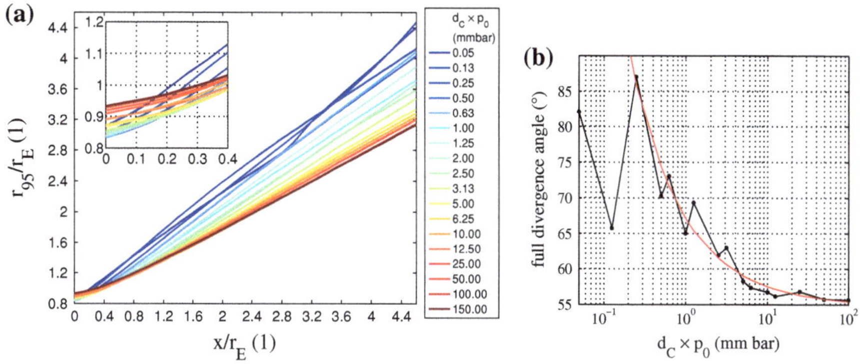

Fig. 3.15 **a** Free jet radius r_{95} containing 95% of the mass flow versus the distance from the nozzle exit for different parameters $d_C \times p_0$. Both axes are normalized to the nozzle exit radius r_E. The *inset* shows the region close to the nozzle exit, **b** Full jet divergence angle versus $d_C \times p_0$. A power law fit is also displayed (*red line*)

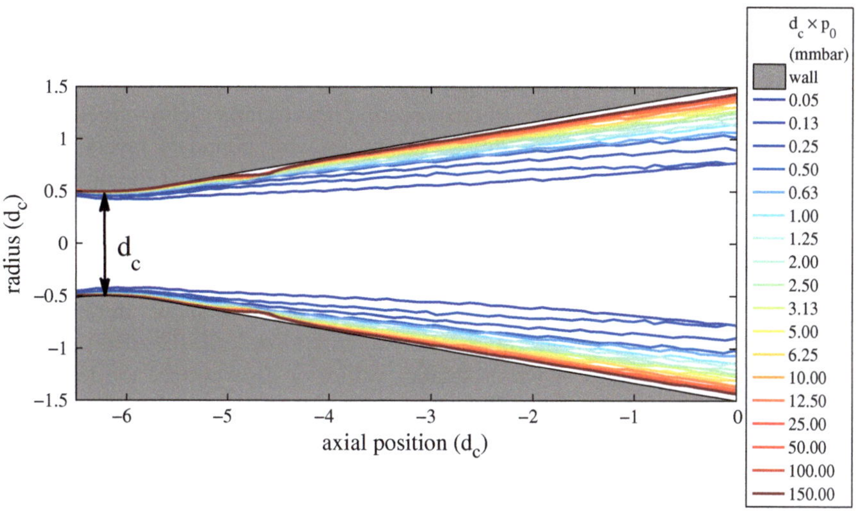

Fig. 3.16 Flow displacement for different parameters $d_C \times p_0$ (mmbar). Both axes are normalized to the critical diameter of each nozzle d_C

to study the development of boundary layers inside the nozzle, again the displacement thickness is studied. Figure 3.16 shows the boundary between the wall bounded flow and the core-flow as it results from calculation of the displacement thickness. Both the x-and the r-axis are normalized to d_C in this graph in order to obtain comparable values. For values of the size-pressure product between 150 and 12 mmbar $(3.1 \times 10^{-4} < Kn < 2.6 \times 10^{-5})$ the change of the displacement

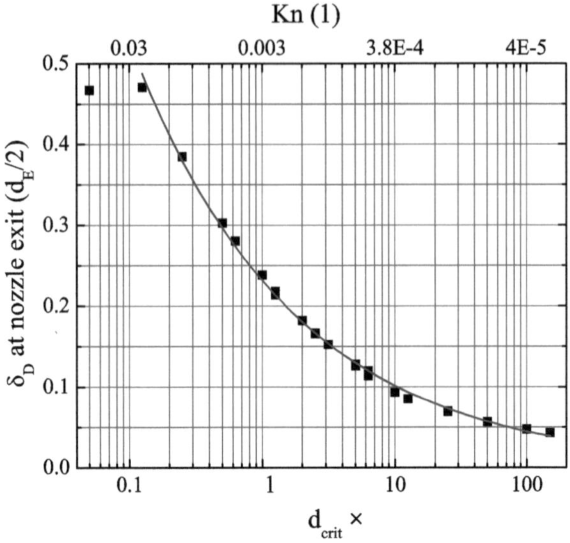

Fig. 3.17 Displacement thickness δ_D at the nozzle exit versus the parameter $d_C \times p_0$, normalized to the exit radius $d_E/2$ of each individual case. The upper *x*-axis shows Kn, calculated at the nozzle exit. The *red line* is a power law fit, see Table 3.4 for details. The off-lying point at $d_C \times p_0 =$ 0.05 mmbar has not been taken into account for the fit because it represents the onset of a different, wall-dominated flow regime

thickness is quite small but then it rises rapidly and finally leads to an effective nozzle contour that hardly expands any more for $d_C \times p_0 = 0.05\,\text{mmbar}(\text{Kn} = 5.3 \times 10^{-2})$. It has been shown in Sect. 3.2.3 that smaller ratios between throat (e.g., critical) cross section and exit cross section lead to larger jet divergence. The modification of the effective nozzle contour by growing boundary layers just has that same effect. Figure 3.17 shows the displacement thickness at the nozzle exit normalized to the exit radius versus $d_C \times p_0$ and versus the Knudsen number Kn, calculated at the nozzle exit. The data can be fitted rather nicely to a simple power law of the form $\delta_D/d_E = a(d_C \times p0)^b$. Here δ_D/d_E is the displacement thickness normalized to the nozzle exit diameter. In Fig. 3.17 this fit is shown in red, the fit parameters are summarized in Table 3.4.

Similarly larger values of Kn or smaller values of $d_C \times p_0$ (see the legend of Fig. 3.18) cause the width of the density gradient at the edge of the jet to increase. The results concerning this effect are shown in Fig. 3.18. It can be seen that sharper gradients at the nozzle exit increase their width more rapidly than smoother ones. The inset shows how the gradients develop as the gas flows along the diverging section of the de Laval nozzle. For the largest values of Kn the gradient width occupies almost the entire nozzle cross section, in which case the flow is dominated by the wall bounded flow and its characteristics are expected (and have been shown) to be significantly distinct from those of low Kn flows generated by large nozzle diameters and/or high pressures. Figure 3.19 shows the variation of the density gradient at the nozzle exit normalized to the exit diameter as a function of $d_C \times p_0$ and Kn. Again, a simple power law of the form $\delta_G/d_E = a(d_C \times p0)^b + c$ represents a quite good fit to the data. δ_G/d_E is the density gradient width normalized to the exit diameter of the nozzle, $a = 0.1549$, $b = -0.3804, c = 0.01073$. Again, the off-lying point at $d_C \times p_0 = 0.05\,\text{mmbar}$

Table 3.3 Parameter study: nozzle size and pressure-list of simulations

Nr. (1)	$\mathbf{d_C}$ (mm)	$\mathbf{d_E}$ (mm)	**L** (mm)	$\mathbf{p_0}$ (bar)	**Kn** (1)
1	1	3	6	0.5	6.0e-3
2	1	3	6	1	3.3e-3
3	1	3	6	2	1.7e-3
4	1	3	6	5	7.4e-4
5	1	3	6	10	3.8e-4
6	1	3	6	25	1.6e-4
7	1	3	6	50	7.8e-5
8	1	3	6	100	4.0e-5
9	1	3	6	150	2.6e-5
10	0.25	0.75	1.5	1	1.1e-2
11	0.25	0.75	1.5	5	2.7e-3
12	0.25	0.75	1.5	10	1.4e-3
13	0.25	0.75	1.5	25	5.9e-4
14	0.25	0.75	1.5	50	3.1e-4
15	0.25	0.75	1.5	100	1.6e-4
16	0.125	0.375	0.75	1	2.1e-2
17	0.125	0.375	0.75	5	5.0e-3
18	0.125	0.375	0.75	10	2.7e-3
19	0.125	0.375	0.75	25	1.2e-3
20	0.125	0.375	0.75	50	5.9e-4
21	0.125	0.375	0.75	100	3.1e-4
22	0.05	0.150	0.3	1	5.3e-2
23	0.05	0.150	0.3	5	1.1e-2
24	0.05	0.150	0.3	10	6.0e-3
25	0.05	0.150	0.3	25	2.7e-3
26	0.05	0.150	0.3	50	1.4e-3
27	0.05	0.150	0.3	100	7.3e-4

Table 3.4 Parameter study: nozzle size and pressure-power law fits

Parameter Name	Fit formula	a	b	c
Displacement thickn.	$\delta_D/d_E = a(d_C \times p0)^b$	0.1163	–0.3718	0
Displacement thickn.	$\delta_D/d_E = a(Kn)^b$	1.218	0.4087	0
Density gradient	$\delta_G/d_E = a(d_C \times p0)^b$	0.1672	–0.3606	0
Density gradient	$\delta_G/d_E = a(Kn)^b$	1.628	0.396	0
Max density	$\rho_{max}/(\rho_0/100) = a(d_C \times p0)^b + c$	2.157	–0.5669	4.212
Max density	$\rho_{max}/(\rho_0/100) = a(Kn)^b + c$	88.14	0.6528	4.278
Mach Number	$M = a(d_C \times p0)^b + c$	–0.7858	–0.4185	4.837
Mach Number	$M = a(Kn)^b + c$	–12.32	0.4882	4.802

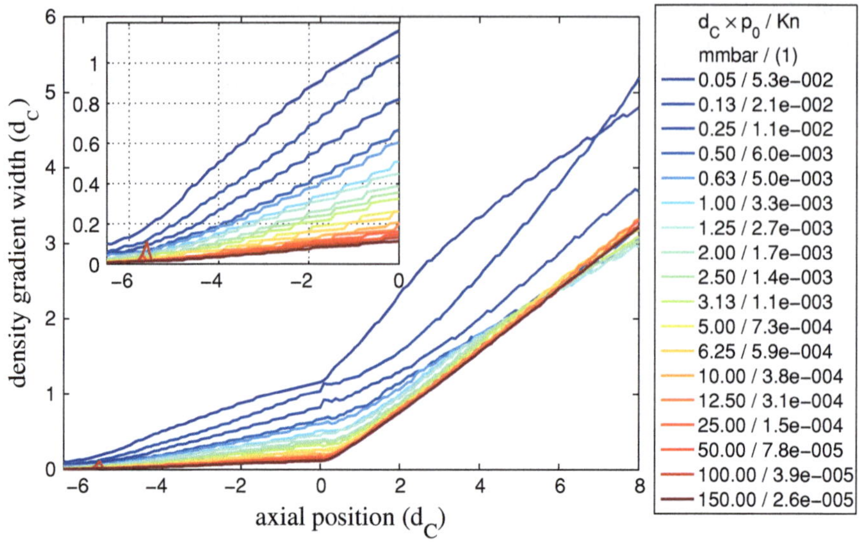

Fig. 3.18 Density gradient width for different parameters $d_C \times p_0$ (mmbar). d_C is the diameter of the nozzle throat, p_0 is the backing pressure. Both axes are normalized to the critical diameter d_C of each individual nozzle

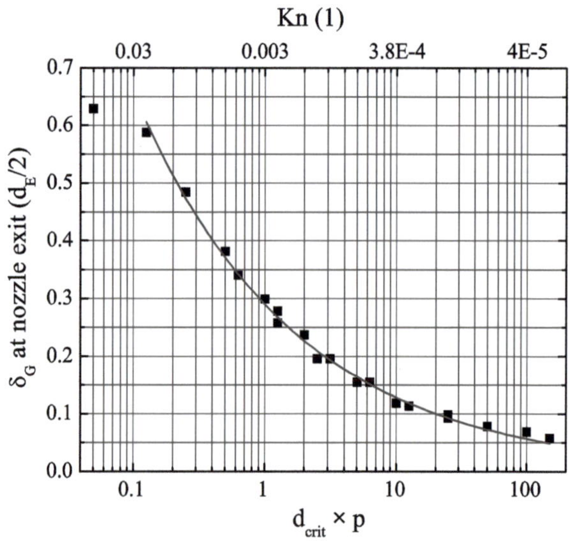

Fig. 3.19 Density gradient width δ_G at the nozzle exit in units of the nozzle exit radius $d_E/2$ versus the parameter $d_C \times p_0$ and Kn. The *red line* is a power law fit, see Table 3.4 for details. The off-lying point at $d_C \times p_0 =$ 0.05 mmbar has not been taken into account for the fit

has not been taken into account for the fit because it represents the onset of a different, wall-dominated flow regime. Obviously, only for values of $d_C \times p_0$ between 150 and 25 can the density gradient be assumed to be approximately independent of the backing pressure and nozzle size.

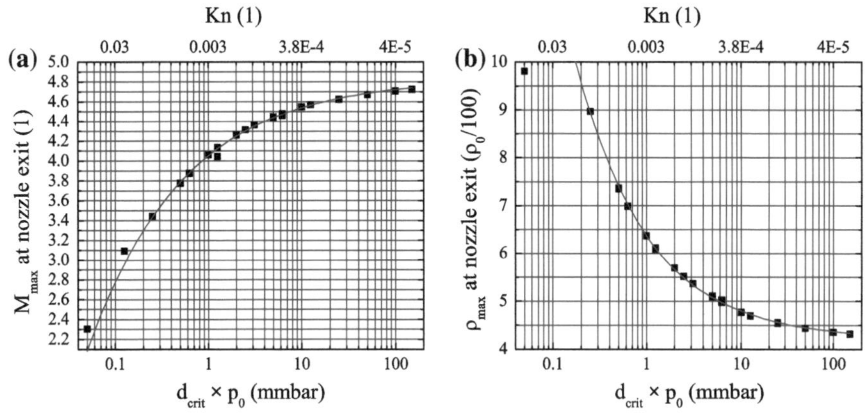

Fig. 3.20 Maximum Mach number and maximum density at nozzle exit versus the parameter $d_C \times p_0$ and Kn. The *red line* is a power law fit, see Table 3.4 for details. The off-lying point at $d_C \times p_0 = 0.05$ mmbar has not been taken into account for the fit *a* Max exit density versus pressure and size. b Max Mach number versus pressure and size

The fact that the variation of the flow parameters with backing pressure and size are caused by the varying thickness of the boundary layers also implies that the Mach number M and the density at the nozzle exit should vary. This is shown in Fig. 3.20. The variation of M with $d_C \times p_0$ and Kn is a direct indication that the effective nozzle cross section changes because the Mach number at the nozzle exit only depends on the reservoir temperature (here always 300 K), the initial velocity (here always zero) and on the ratio between the critical and the exit cross section. Figure 3.20b shows that also the density at the nozzle exit, measured in percent of the reservoir density, is not constant. This has two important consequences: first, for the same backing pressure a smaller nozzle will give higher density than smaller nozzles. And second, the density at the nozzle exit is not linearly proportional to the backing pressure but has a power-law dependence on it. However, Fig. 3.21 shows that in a wide parameter range the error introduced by assuming linear dependence is small enough to allow this approximation to be used in many experiments.

Finally, as in the section before, the flatness of the density line-outs inside the nozzle is analyzed. The results are presented in Fig. 3.22a which shows density line-outs at the nozzle exit. For the cases with low Knudsen number Kn, there is a distortion close to the nozzle wall. The amplitude of this distortion shrinks as Kn rises, yielding rather smooth flat-top profiles for values of Kn between 5×10^{-4} and 2×10^{-3} albeit with ever broader gradient width. For even smaller values of Kn, the profiles completely loose their flat-top character as the flow becomes more and more dominated by wall effects. This is also reflected in plot 3.22b where it can be seen that the peak-to-peak fluctuation in the core flow shrinks as Kn rises. For the largest values of Kn, essentially no core flow is left which is why also the peak-to-peak variation of it drops to zero.

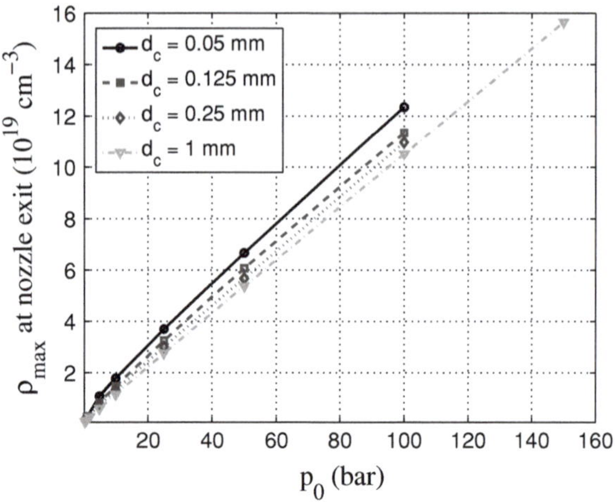

Fig. 3.21 Maximum density ρ_{max} at the nozzle exit versus backing pressure p_0. An approximately linear behavior can be assumed in a wide parameter range

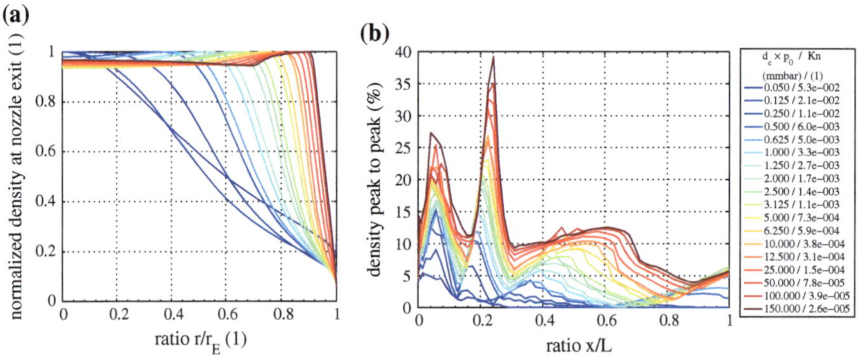

Fig. 3.22 Comparison of density line-outs along lines of constant x at the nozzle exit (**a**) and relative density peak-to-peak variation along the diverging section of the nozzles (**b**). All plots in (**a**) are scaled to the same size in order to be easily comparable. The x-positions in (**b**) are normalized to the length L of the diverging nozzle section. The legend of (**b**) is valid for both figures. **a** Normalized density line-outs at nozzle exit, **b** relative density peak-to-peak fluctuation of the core flow

3.2.5 Effects of Non-Negligible Background Pressure

In real experimental conditions, the gas jet does not emanate into a perfect vacuum but encounters a certain background pressure in the surrounding space. In fact, the gas load constituted by the jet itself usually gives the largest contribution to the development of a certain background pressure in the vacuum chamber. To check whether for realistic experimental conditions these variations in background pressure lead to a significant distortion of the jet, a study with a de Laval nozzle having 1 mm throat diameter, 3 mm exit diameter, a length of the diverging nozzle

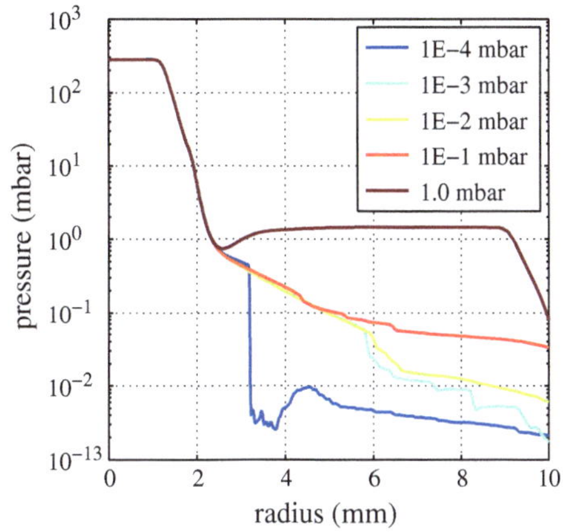

Fig. 3.23 Density profiles 1 mm after the exit of a de Laval nozzle with 1mm throat diameter, 3mm exit diameter and a length of the diverging nozzle section of 12 mm. The backing pressure is 50 bar, Helium. The background pressure varies between 10^{-4} and 1 mbar

section of 12 mm was conducted. The backing pressure is 50 bar, the medium is Helium, and the background pressure is varied between 10^{-4} and 1 mbar. Figure 3.23 shows line-outs of density at a distance of 1 mm from the nozzle exit for all the simulated cases. It can be seen that—as it is expected—the main gas jet remains unaffected by the varying background pressure. Only the very low pressure gas plume that surrounds the gas jet is influenced by the backing pressure at all. This behavior only changes when the background pressure lies within the same order of magnitude of the static pressure in the gas jet at the nozzle exit which lies around 300 mbar in this case.

3.2.6 Gas Targets with Additional Degrees of Freedom

In many experiments it is desirable to dispose of a gas target with a somehow structured density profile. Here, a short summary is given about two such gas targets that were considered in this study. The first target consists of two de Laval nozzles that are close to each other and thereby generate a double-plateau like density profile. A sketch of the setup together with a density contour plot is shown in Fig. 3.24. The simulation is 2D, the medium is Helium. The small nozzle has a 50 μm throat and a 100 μm exit, the backing pressure is 50 bar. The large nozzle has a throat of 100 μm and an exit of 400 μm and a backing pressure of 10 bar. The separation between the two nozzle exits is 10 μm. Outside of the nozzles a structure of shocks develops in the region where the two gas jets touch each other. However, in the region close to the nozzle exits—which is the most interesting one for laser–plasma experiments that usually require steep gradients and high density—there is a region where the two jets do not disturb each other yet and form two density plateaus with a small dip in between. By controlling the backing pressure of the two nozzles, the densities of the two plateaus can easily be optimized during experiments. By separating the two nozzles and mounting them on

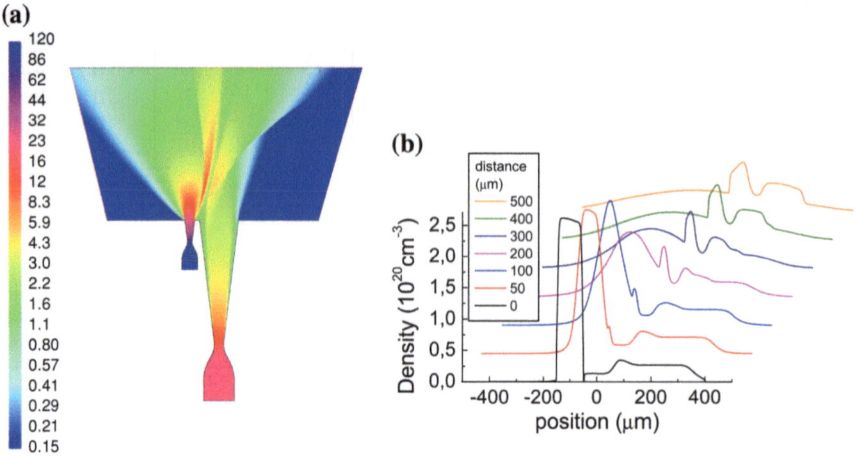

Fig. 3.24 Density contours **a** and horizontal density line-outs **b** of a double nozzle consisting of two de Laval nozzles. Data from 2D-simulation. Density in **a** is given in (cm^{-3})

translation stages, it would also be possible to adjust the separation between the two jets as an additional degree of freedom.

Another interesting option for designing more complex gas targets is offered by the exploitation of shock fronts in the supersonic gas jet. Shock fronts are easily obtained by simply putting an obstacle in the way of the jet, for example in the form of a sharp edge. It has been detailed in Sect. 2.1.3 how shock fronts develop in supersonic flows, Eqs. 2.36–2.40 giving all important relations for normal and oblique shocks. It can be seen there that the ratios of the flow parameters immediately before and after the shock only depend on the initial Mach number of the gas jet and on the angle between the initial flow direction and the shock front.

Figure 3.25 shows the 2D-simulation results. The width of the depicted two dimensional gas channel is 300 μm, the length of it is 600 μm. As can be seen in the contour plots, this length is just sufficient to prevent any flow distortions caused by the plate in the jet to propagate back to the input boundary. This allows in turn to set the input boundary to (known) undisturbed flow conditions. With a distance of 50 μm from the exit of the channel, a 5 μm thick plate is placed that represents a razor blade that is put into the supersonic jet. The overlap with the jet is 7.5 μm which—as can be seen in Fig. 3.25a—is just enough to penetrate through the entire subsonic boundary layer and block a small fraction of the supersonic core flow. Relations (1.36–1.40) dictate how the shock front changes qualitatively as the knife edge is moved in and out of the beam: as the knife edge is moved more and more into the beam, a larger part of the jet is blocked, causing an ever stronger shock at increasing angles with respect to the flow direction. Moving the edge out, the angle and the strength of the shock decrease. Once the knife edge does not penetrate the core flow any more but only blocks a part of the subsonic boundary layer, the shock decays into Mach waves that only constitute minor distortions of

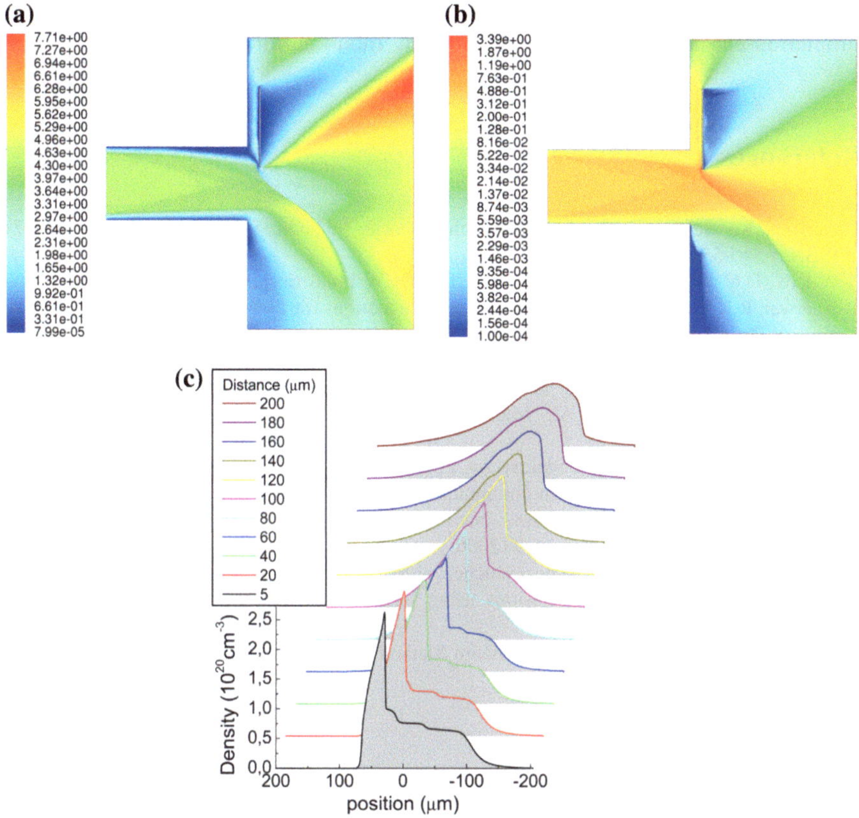

Fig. 3.25 **a** shows contours of Mach number, **b** contours of density $(\mathrm{kg/cm^3})$ and **c** line-outs of density $(\mathrm{particles/cm^3})$ along lines perpendicular to the flow axis at different distances from the knife edge. The flow at the input boundary is initialized with the exit conditions of a de Laval nozzle, providing a Mach number of 4.5. A small solid-plate is introduced into the gas jet in order to generate a shock in the jet

the flow. Figure 3.25b shows the density contours of the flow. The sharp increase in density within the shock front can easily be seen. Figure 3.25c shows line-outs of particle density along vertical lines at different distances after the knife edge. The density ratio across the shock is larger than 3 very close to the edge and then drops steadily to about 2.4 at a distance of 100 μm from the knife edge. As expected, this goes along with a decreasing angle between the shock and the flow direction. This angle is 45° in a distance of 5 μm from the knife edge and then decreases to a minimum of 31° degrees before it then grows again as the shock enters the regions with smaller Mach numbers. By evaluating Eq. 2.36 with these numbers, e.g. substituting $M sin\gamma$ for M with M = 4.5, $\gamma = 45° - 31°$, one obtains a density ratio of 3.1 for 45° and one of 2.6 for 31°, thus, being in good agreement with the simulation results. Since the maximum value $(\kappa + 1)/(\kappa - 1)$ (see Sect. 2.1.3) for the density ratio which is asymptotically reached for $M \rightarrow \infty$ is

4.0 it is clear that no much stronger density jumps than the one in the present simulation can actually be reached with any mono-atomic gas.

References

1. Gersten, K., Schlichting, H.: Grenzschicht-Theorie, 9th edn. Springer Heidelberg (1997). ISBN 3-540-55744-X
2. Wilcox, D.C.: Turbulence Modeling for CFD. DCW Industries, 3rd edn. (2006)
3. Agarwal, R.K.: Beyond Navier–Stokes: Burnett equations for flows in the continuumtransition regime. Phys. Fluids **13**(10):3061–3085(2001)
4. Kujawa, J., Hitt, D.L.: Transient shutdown simulations of a realistic mems supersonic nozzle, pp. 2004–3762. AIAA (2004)
5. Karniadakis, G.E., Besok, A.: Micro Flows, Fundamentals and Simulation. Springer, Heidelberg (2002)
6. Mo, H., Lin, C., Gokaltun, S., Skudarnov, P.V.: Numerical study of axisymmetric gas flow in conical micronozzles by dsmc an continuum methods, pp. 2006–991. AIAA (2006)
7. Gadepalli, V.V.V., Chengxian, lIn.: Navier–Stokes modeling of gas flows in a delaval micronozzle, pp. 2006–1425. AIAA (2006)
8. Tejeda, G., Maté, B., Fernández-Sánchez, J.M., Montero, S.: Temperature and density mapping of supersonic jet expansions using linear Raman spectroscopy. Phys. Rev. Lett. **76**(1):34–37(1996)
9. Miller, D.R.: Atomic and Molecular Beam Methods, vol. 1. Oxford University Press, Oxford (1988)

Chapter 4
Experimental Characterization of Gas Jets

4.1 Experimental Setup for Characterizing Gas Jets

Several different gas jet designs were realized and tested in order to verify their functionality for the experiments. Drawings of these designs can be found in appendix B. The nozzles all have an angle of approximately 7° and a ratio between the throat and the exit diameter of 3. This seems to be an adequate compromise between maximum reachable density, jet collimation, gradient steepness, and quality of the flat top profile. Manufacturing these nozzles with exit diameters below 1.5 mm proved to be a veritable challenge, in the end only by electro-erosion could those nozzles be constructed. The characterization of the nozzles was done by use of a Mach–Zehnder interferometer, see Fig. 4.1. The nozzle was mounted into a vacuum chamber that permitted measurements under conditions comparable to those in the laser–plasma experiments. In order to be able to maintain a pressure below 10^{-2} mbar, the nozzles were operated in pulsed mode. To this end, commercial electromagnet valves (Parker Hannifin, valve 9S1-A1-P4-9B04 and IOTA One driver) were directly attached to the $1/2''28$ UNEF thread of the nozzle heads, see drawings in appendix B, Fig. B.1–7. Rise and fall time of these valves were measured to be up to 500 μs, so a total opening time of 1.2 ms was used in order to ensure full opening of the nozzle and reproducible conditions from shot to shot. Although the laser–plasma experiments were usually conducted using Helium, in the characterization Argon was employed due to its larger refractive index. Table 4.1 displays the refractive indices of some gases at normal conditions (1.01325 bar, 273 K) for a wavelength of 532 nm together with the phase shift that is imprinted onto a light ray of that wavelength that passes through 1 mm of the gas at a density of $10^{19}\,\mathrm{cm}^{-3}$. Values are from [1]. Since the exit velocity of the gas is 1.6 km/s for Helium and 526 m/s for Argon, a pulsed laser was used for the interferometry. The short pulse duration (20 ps) ensured imaging without velocity-smearing of flow features.

As an additional diagnostic, a fast pressure transducer was connected to the high pressure side of the pulsed valve. This allowed to detect unwanted reflections

K. Schmid, *Laser Wakefield Electron Acceleration*, Springer Theses,
DOI: 10.1007/978-3-642-19950-9_4, © Springer-Verlag Berlin Heidelberg 2011

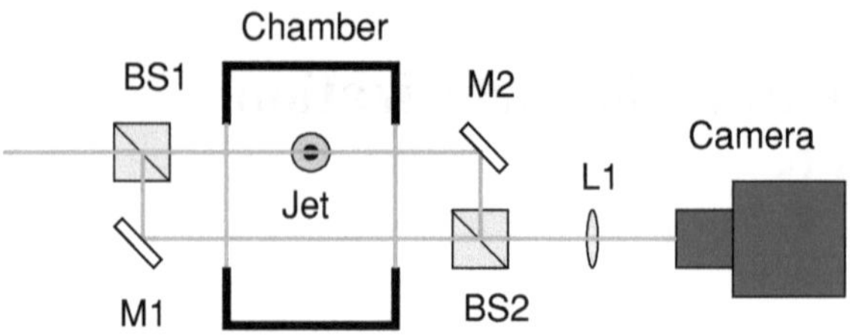

Fig. 4.1 Sketch of the experimental setup used for characterizing the gas jets. Both arms of a Mach–Zehnder interferometer (BS1, BS2, M1, M2) are passed through a vacuum chamber, one of them also traverses the pulsed gas jet that resides inside the chamber. A Lens (L1) is used to image the gas jet onto a CCD camera. The laser is a Nd:YLF amplifier producing pulses at 10 Hz, with 300 μJ per pulse and a pulse duration of 20 ps

Table 4.1 Refractive index and phase shift relative to vacuum for several gases

Gas type (1)	(n−1) (1)	$\Delta\phi$ (π rad)
He	3.5×10^{-5}	0.05
H2	1.4×10^{-4}	0.21
Ar	2.8×10^{-4}	0.41
N2	2.85×10^{-4}	0.42

The phase shift is calculated for a wavelength of 532 nm passing through 1 mm gas with a density of $10^{19}\,\text{cm}^{-3}$.

of the pressure-wave caused by the pulsed valve and also turned out to be very useful for tuning the valve.

4.2 Numerical Evaluation of Experimental Data

In order to get back the density of a gas-slab of given length from the phase shift it imprints onto a light ray traversing it, a connection between the gas density and its refractive index must be established. Measured values of the refractive index of many gases under normal conditions (1.01325 bar, 273 K) can be found, for example, in [1]. The well-known Clausius–Mosotti relations [2]

$$\frac{n^2-1}{n^2+2} = \frac{\rho_P}{3}\gamma_{mol} \tag{4.1}$$

can now be used to find this connection. Here, n is the refractive index, ρ_P is the particle density (cm^{-3}), and γ_{mol} is the molecular polarizability of the gas. With the approximation $(n^2-1)/(n^2+2) \approx 2/3(n-1)$ this can be written in the form

$$n - 1 = K\rho_P \tag{4.2}$$

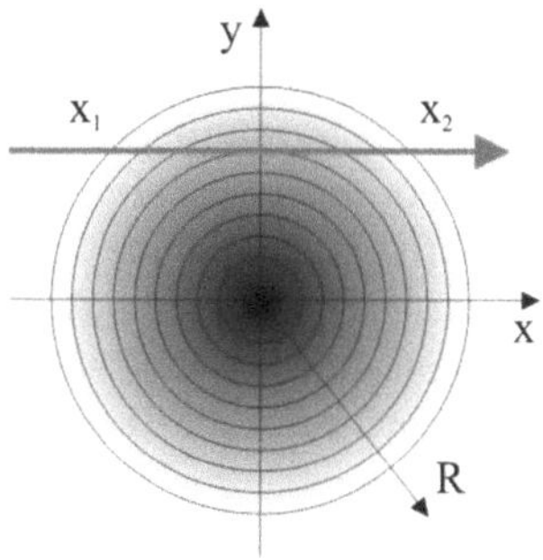

Fig. 4.2 Abel inversion. A laser beam (*green*) passes a phase object (*grey shades*) thereby accumulating a phase shift. Under the assumption of axial symmetry, the Abel inversion allows to retrieve the radial distribution of the refractive index of the phase object from the accumulated phases at all y positions

This is known as the Gladstone–Dale relationship and is very well fulfilled for gases. The constant K can now be calculated for each gas with one known value pair of n and ρ_P.

Since the laser beam samples the whole jet only from one direction, the accumulated phase shift corresponds to a projection of the gas density along the direction of propagation of the beam. This can bee seen in Fig. 4.2 where the laser (green arrow) passes the phase object, e.g. the gas jet, from left to right. Therefore, the difference between its accumulated phase shift and the one of the beam passing through vacuum $\Delta\Phi$ has the form

$$\Delta\Phi(y) = \frac{4\pi}{\lambda} \int_{y}^{r_0} \frac{(n(r) - 1)r}{\sqrt{r^2 - y^2}} dr \tag{4.3}$$

Here, r_0 is a radius sufficiently large to be able to neglect any phase contributions from regions outside that radius. Under the assumption that the phase object is axially symmetric—which is certainly the case for a round gas jet—this equation can be inverted using the well-known Abel inversion:

$$\frac{2\pi}{\lambda}(n(r) - 1) = -\frac{1}{\pi} \int_{r}^{r_0} \frac{d}{dy}(\Delta\Phi(y)) \frac{dy}{\sqrt{y^2 - r^2}} \tag{4.4}$$

and allows to get back $n(r)$ and with Eq. 4.2 $\rho_P(r)$ from the measured projected phase shifts $\Delta\Phi(y)$.

The fact that the derivative of the measured signal $\Delta\Phi$ appears in the above formula leads to a strong sensitivity of the result of the inversion to noise in the original data. Therefore, care must be taken when it comes to actually numerically inverting the measured phase shifts [3]. In this work, the numeric Abel inversion was done using the free tool IDEA available from TU Graz at http://optics.tu-graz.ac.at/idea/idea.html.

In order to remove unwanted features like interference patterns from dust particles from the measured interferograms, the data was low-pass filtered using a fast fourier transform (FFT) filter. The resulting two-dimensional phase data was then Abel-inverted using IDEA. For inversion, the fourier transform method [4, 5] and the Backus–Gilbert method [6, 7] yielded the best results. In most cases, the Backus–Gilbert method was used. The numeric Abel-inversion gives back the right-hand side of Eq. 4.4 in units of (1/pixel) where *pixel* is one pixel of the acquired image. So, in order to obtain the actual density profile, the result of the inversion must be converted to (particles/cm^3) by multiplication with a scaling factor S_{ai}

$$S_{ai} = \frac{\lambda}{2\pi\xi K} \tag{4.5}$$

where $K\,(\mathrm{cm}^3)$ is the Gladstone–Dale constant (4.2), $\lambda\,(\mu\mathrm{m})$ is the laser wavelength and $\xi\,(\mu\mathrm{m}/\mathrm{pixel})$ is the scale of the acquired image. Multiplication of the Abel-inverted phase shift with S_{ai} yields the density in units of $(10^{19}\ \mathrm{cm}^{-3})$. For Argon this evaluates to $S_{ai} = 1.44845 \times 10^{22} \times \lambda/\xi$.

4.3 Experimental Results on Gas Jets

As mentioned above, one of the main goals was to establish whether supersonic gas nozzles with sizes in the 100 μm regime would produce collimated, high density gas jets suitable for laser–plasma experiments. The smallest nozzle design used in this study (and also applied in the experiments) has a throat diameter of 50 μm, an exit diameter of 150 μm, and a length of the diverging section of 400 μm. A drawing of this nozzle can be found in Appendix B, Fig. B.1. Several of these nozzles were produced, showing sometimes—due to the delicate manufacturing process—quite substantial deviations from the design goals. In the following, measurement results of such a nozzle are presented. By microscopy, an elliptic throat of 92 × 80 μm diameter, a nozzle exit of 178 × 174 μm and a length of the diverging section of the nozzle of 400 μm were determined. Results of density measurements using Argon are shown in Fig. 4.3. Figure 4.3a shows the phase shift caused by the Argon gas jet produced by the nozzle. The phase was retrieved from the interferometric measurement and subsequent data analysis detailed above. Shown are three different shots (thin black lines) and their average (thick black line). The lower panel shows the relative deviation of the three shots from the average. The oscillations with a period of roughly 30 μm are measurement artefacts generated during data processing, so a real shot to shot reproducibility of better than 2 to 3% can be assumed. The dashed orange line shows the phase shift calculated from the results of simulations that were conducted for this specific case. The agreement is

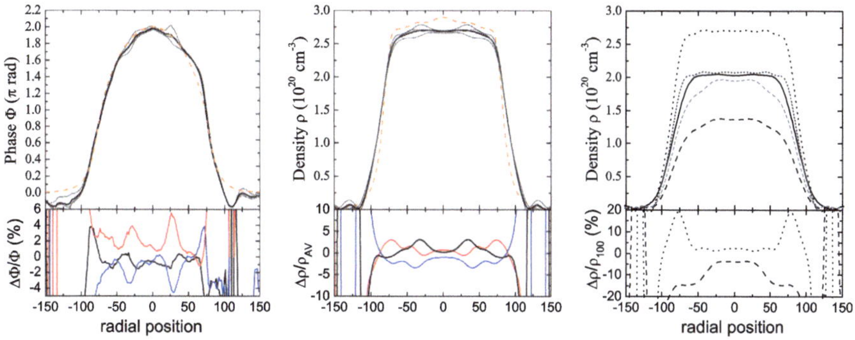

(**a**) Mesured phase shift 130 bar. (**b**) Retrieved density 130 bar. (**c**) Backing pressure variation.

Fig. 4.3 Measured density profiles of an Argon gas jet produced by a de Laval nozzle with 86 μm throat and 176 μm exit diameter. All profiles are located at a distance of 50 μm from the nozzle exit. The *upper panel* in **a** depicts the measured phase of three different shots (*thin black line*) and their average (*thick black line*) and calculated phase shift from simulation result of that nozzle (*dashed orange line*). The *lower panel* shows the relative deviation of each of the three shots from the average showing the typical value of 5%. The *upper panel* of **b** shows results of Abel-transformations of the measured phase data in (**a**) together with the corresponding simulation result. Line types and colors are as in (**a**). **c** Shows three shots at 70 (*dashed*), 100 (*solid*) and 130 bar (*dots*) backing pressure, respectively. The *thin grey lines* show the density profiles for 70 bar and 130 bar scaled linearly to 100 bar backing pressure. The *lower panel* shows the deviation of the scaled profiles from the result at 100 bar of approximately 5%. The gradient at the jet-edge steepens for higher backing pressures

very good although perfect agreement cannot be expected because the simulations are done in axisymmetric geometry, so the elliptic nozzle cross sections had to be approximated by average circular cross sections. Figure 4.3b shows Abel inversions of the retrieved density profiles. The gas jet is flat-top with a gradient width defined as a density rise from 10 to 90% of 35 μm and a 142 μm flat top. Similar as for the phase data, the shot to shot fluctuations of the density profiles is less than 2 to 3%. Figure 4.3c shows retrieved density profiles for varying backing pressures of 70 (dashed) 100 (solid) and 130 (dots) bar. The thin grey lines are scaled versions of the 70 and 130 bar profiles. Scaling was done assuming a linear dependence of the jet density on the backing pressure. Small but significant deviations from linearity can be seen and are plotted in the lower panel as the relative deviation of the scaled density profiles from the 100 bar profile. This qualitatively reproduces the scaling predicted by simulations, see Sect. 3.2.4. Quantitative correspondence cannot be expected due the different nozzle geometry used in the simulations. Figure 4.3c also shows that the gradient width at the edge of the gas jet rises with dropping pressure which again is in qualitative agreement with the simulations.

As a side-remark, it is noted that the distance of 50 μm, for which the above results are given, is approximately one third of the nozzle exit diameter. The microscopic de Laval nozzle is able to produce a flat top profile with steep

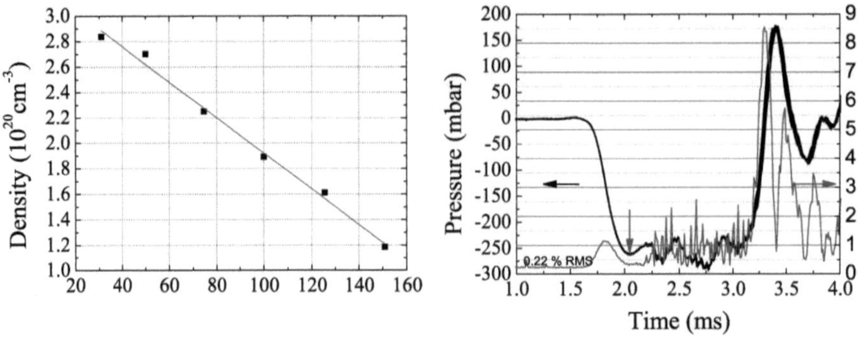

(a) Variation of density with distance from the nozzle exit.

(b) Shot-to-shot pressure variation on the high pressure side of the nozzle.

Fig. 4.4 Figure **a** shows on axis density versus distance from nozzle. The red line is a linear fit to the data. Figure **b** shows the signal of a fast pressure transducer attached to the high pressure gas feed of the nozzle. 100 shots are shown. The electronic noise on the signal is 0.22% RMS, the RMS—fluctuation of the signal at a delay of 2.1 ms (red arrow) is 0.5%. If used at this delay, this should also be representative for the shot-to-shot reproducibility of the gas jet produced by the nozzle

gradients at this distance. A cylindrical nozzle of this size—which would be much easier to produce—would yield a nearly gaussian profile at this distance (Sect. 2.2). However, the distance of 50 μm is more or less the minimum distance to which a high power laser can still be focused without destroying the nozzle tip. Therefore, microscopic de Laval nozzle are the only viable option of producing small flat top targets at useful working distances.

Measurement results concerning the dependence of the gas jet density on the distance from the nozzle exit are shown in Fig. 4.4a. It depicts the measured on-axis density versus the distance from the nozzle exit. Within the considered range from the nozzle exit, which is limited by the signal to noise ratio of the phase shift measurement, the gas jet density drops linearly with increasing distance from the nozzle. In order to obtain more accurate information about the shot-to-shot reproducibility than is possible by interferometry—the inherent accuracy of this is approximately 5%—a fast pressure transducer was connected to the gas feed of the nozzle. The recorded signal is shown in Fig. 4.4b. The drop of the signal to roughly −1 V between 1.7 and 3.3 ms corresponds to the pressure drop in the gas feed caused by the opening of the nozzle. The signal shows a smooth opening of the nozzle between 1.7 and 2 ms followed by an almost constant stretch between 2 and 2.1 ms. Within this time window, the RMS-fluctuation of the signal over the 100 shots acquired in the measurement is approximately 0.5%. If used at this delay, this number should also be representative for the shot-to-shot reproducibility of the gas jet produced by the nozzle. During the remaining nozzle opening time from 2.1 to 3.1 ms, some oscillations occur in the gas duct, in this region the signal-RMS rises to approximately 1 to 2%. And finally the nozzle closes between 3.1 and 3.3 ms. The nozzle driver was set to an opening time of 1.2 ms during this measurement.

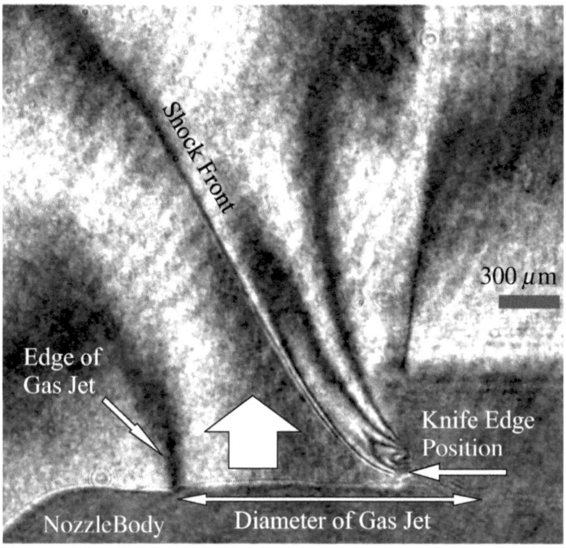

Fig. 4.5 Interferometric visualization of the shock front generated by introducing a knife edge into the supersonic gas jet emanating from a de Laval nozzle. The nozzle has an exit diameter of 1.5 mm, the backing pressure is 50 bar, the medium is argon. The interferometric image was obtained with the experimental setup detailed in Sect. 4.1. In this case, the interferometer was aligned such that the phase fronts of the two arms are parallel to each other in the imaging plane. This is done by maximizing the fringe-separation in the interferogram up to the point where the distance between to fringes is larger than the field of view of the camera recording the interferogram, so that ideally the entire image would show only one shade of grey. Then each phase-modulation introduced by a phase-object as the gas jet or the shock front shows up in the image as grey-scale contours proportional to the phase shift, see for example [8]. The shock front has an angle of approximately 60° with respect to the gas flow direction. For this angle, theory predicts a density jump by a factor of 3.4 (see text for details)

4.4 Shock Fronts in Supersonic Gas Jets

In Sect. 3.2.6, the production of a very steep density gradient inside a gas target via supersonic shocks was proposed. Here, first experimental results are presented. The experimental setup consists of a pulsed de Laval nozzle with an exit diameter of 1.5 mm (see appendix B, Fig. B.6) that is equipped with a knife edge which is introduced from one side into the gas jet, thereby blocking one part of it. The knife edge causes a shock front that traverses the jet. Due to the rather large density jump in the shock, it is easily detected via interferometry. Figure 4.5 shows the corresponding image. The interferometer is adjusted such that it produces no fringes without gas. This so-called infinite fringe width alignment is achieved by reducing the angle between the interfering wave fronts to zero. In this way, any phase shifts picked up by one of the interferometer arms shows up as an intensity modulation in the image. These images have the advantage of being easily readable by the human eye but cannot be quantitatively analyzed with sufficient accuracy. However, the axial symmetry is broken by the shock front so quantitative analysis using Abel inversion cannot be used any more in any case. Therefore, the infinite fringe width alignment was chosen here for flow visualisation, leaving the quantitative characterization by tomography for future work. Figure 4.5 shows the measured interferogram produced with a backing pressure of 50 bar and Argon. The gas jets emanates from the nozzle in vertical direction as indicated by the large arrow. The knife edge is introduced to the jet from the right, it overlaps with it by 320 μm, at a distance of approximately 50 μm from the nozzle exit. The knife edge is a commercial razor blade of 100 μm thickness and is slightly tilted with respect to the laser beam so that it appears thicker than that in the image.

The shock front caused by the knife edge is almost perpendicular to the gas jet as it starts at the knife edge and then gradually bends upwards to reach an angle of approximately 60° which then stays constant throughout the core of the jet. In the upper left corner of the image it then starts to penetrate the density down-ramp at the opposite edge of the jet, bending downwards again. This behavior is nicely reproduced in the simulations presented in Sect. 3.2.6, Fig. 3.25. In Sect. 2.1.3, the analytic treatment of supersonic shocks is summarized yielding the main result that the ratio of all state variables before and after the shock depend only on the Mach number M of the gas jet and on the angle of the shock. Since M cannot be measured easily, it must either be calculated [(see Sect. 2.1.4, Eq. 2.58] analytically or obtained via simulation. For Argon no simulation has been done for this case, so the analytical value of M = 4.83 is taken. Since the strength of the shock starts to saturate around M = 6 (see Fig. 2.2), the error introduced by any uncertainty in M is small. With M = 4.8 and an angle of $\gamma = 60°$, an effective Mach number of $Msin(\gamma) = 4.19$ is obtained. By substituting this for M in Eq. 2.36, a density rise in the shock by a factor of 3.42 is obtained. The corresponding values for Helium are equal as by the one-dimensional analytical theory. A simulation with Helium predicts a slightly lower exit Mach number of 4.64. This then leads to a density rise by a factor of 3.37 which is practically the same as the analytical result.

The resolution of the interferometric measurement at the moment does not permit determination of the width of the shock, so the statement that it is on the order of the molecular mean free path, which is around 500 nm in this case, must suffice at this point. This is on the same order of magnitude as the typical laser wavelength of 800 nm, and at least one order of magnitude smaller than the typical plasma wavelength. This close-to-ideal situation might be very interesting for future laser–plasma experiments.

References

1. Nikogoyan.: Properties of Optical and Laser-Related Materials. Wiley, New York
2. Jackson, J.D.: Classical Electrodynamics, 3rd edn. Walter de Gruyter, Berlin (2002)
3. Pretzler, G., Jäger, H., Philipp, H., Woisetschläger, J.: Comparison of different methods of abel inversion using computer simulated and experimental side-on data. Z. Naturforsch **47**a, 955 (1992)

4. Kalal, M., Nugent, K.A.: Abel inversion using fast fourier transforms. Appl. Optics **27**(10), 1956 (1988)
5. Pretzler, G.: A new method for numerical abel-inversion. Z. Naturforsch **46**a, 639 (1991)
6. Backus, G.E., Backus, G.E., Backus, G.E.: The resolving power of gross earth data. Geophys. J. R. Astron. Soc **16**, 169 (1968)
7. Backus, G.E. et al.: Uniqueness in the inversion of inaccurate gross earth data. Phil. Trans. R. Soc. A **266**, 123 (1970)
8. Merzkirch, W.: Flow Visualization. Academic Press, New York (1987)

Part II
Few-Cycle Laser-Driven Electron Acceleration

Chapter 5
Electron Acceleration by Few-Cycle Laser Pulses: Theory and Simulation

5.1 Introduction to Relativistic Laser–Plasma Physics

5.1.1 Non-Relativistic Cold Collisionless Plasmas

In the following, the plasma is treated as an electron fluid, the ions serve only as a fixed neutralizing background. This approximation is justified for all processes on timescales much shorter than the inverse ion–plasma frequency which will be derived below. In order to derive the dispersion relations of a cold collisionless plasma we start from the Maxwell equations:

$$\nabla \times \mathbf{E} = -\frac{1}{c}\frac{\partial}{\partial t}\mathbf{B} \tag{5.1}$$

$$\nabla \cdot \mathbf{E} = 4\pi\rho \tag{5.2}$$

$$\nabla \times \mathbf{B} = \frac{1}{c}\frac{\partial}{\partial t}\mathbf{E} + \frac{4\pi}{c}\mathbf{j} \tag{5.3}$$

$$\nabla \cdot \mathbf{B} = 0 \tag{5.4}$$

Here, $\mathbf{E}$ is the electric field, $\mathbf{B}$ the magnetic field, ρ the charge density and $\mathbf{j}$ the current density, all are functions of position $\mathbf{r}$ and time t. c is the vacuum speed of light. Since in a plasma electromagnetic as well as electrostatic (charge separation) waves exist, it makes sense to split the equations into a electromagnetic part containing only the part of the electric field that has a curl and into an electrostatic part where the electric field has no curl and can, therefore, be expressed as the gradient of a scalar potential:

$$\mathbf{E} = \mathbf{E}_{rot} + \mathbf{E}_{grad} = -\frac{1}{c}\frac{\partial}{\partial t}\mathbf{A} - \nabla\Phi \tag{5.5}$$

K. Schmid, *Laser Wakefield Electron Acceleration*, Springer Theses,
DOI: 10.1007/978-3-642-19950-9_5,

Here, **A** is a purely rotational vector potential and Φ is a scalar potential. The B field of course is purely rotational and can be expressed by

$$\mathbf{B} = \nabla \times \mathbf{A} \tag{5.6}$$

The electromagnetic (rotational) equations then are given by

$$\nabla \times \mathbf{E}_{rot} = -\frac{1}{c}\frac{\partial}{\partial t}\mathbf{B} \tag{5.7}$$

$$\nabla \cdot \mathbf{E}_{rot} = 0 \tag{5.8}$$

$$\nabla \times \mathbf{B} = \frac{1}{c}\frac{\partial}{\partial t}\mathbf{E}_{rot} + \frac{4\pi}{c}\mathbf{j}_{rot} \tag{5.9}$$

$$\nabla \cdot \mathbf{B} = 0 \tag{5.10}$$

and the electrostatic ones by

$$\nabla \cdot \mathbf{E}_{grad} = 4\pi\rho \tag{5.11}$$

$$-\frac{1}{c}\frac{\partial}{\partial t}\mathbf{E}_{grad} + \frac{4\pi}{c}\mathbf{j}_{grad} = 0 \tag{5.12}$$

In order to obtain a dispersion relation from the electromagnetic set (5.7–5.10) a relationship between the current and the electrical field is needed. In the linear approximation, so by neglecting all terms of order $|E|^2$ the linearized equation of motion reads

$$\frac{\partial}{\partial t}\mathbf{u} = -\frac{e}{m_e}\mathbf{E}(x)\exp(-i\omega t) \tag{5.13}$$

with **u** being the velocity of the electron fluid and $\mathbf{E} = \mathbf{E}(x)\exp(-i\omega t)$ the laser electric field. With the current density being determined by

$$\frac{\partial}{\partial t}\mathbf{j}_{rot} = -N_0 e\frac{\partial \mathbf{u}}{\partial t} = \frac{\omega_p^2}{4\pi}\mathbf{E} \tag{5.14}$$

and the plasma frequency

$$\omega_p^2 = \frac{4\pi e^2 N_0}{m} \tag{5.15}$$

with N_0 being the electron density one obtains for the current

$$\mathbf{j}_{rot} = \frac{i\omega_p^2}{4\pi\omega}\mathbf{E} \tag{5.16}$$

Assuming linearly polarized plane waves propagating in x direction, the vector potential $\mathbf{A}$ is given as $\mathbf{A} = Re\{\mathbf{A}_0 e^{i\psi}\}, \mathbf{A} = A_0 \hat{e}_y$ and with the phase $\psi = \mathbf{kr} - \omega t$ one obtains

$$\mathbf{E}_{rot} = -\frac{1}{c}\frac{\partial}{\partial t}\mathbf{A} = Re\left\{\frac{i\omega}{c}\mathbf{A}_0 e^{i\psi}\right\} \tag{5.17}$$

$$\mathbf{B} = \nabla \times \mathbf{A} = Re\{i\mathbf{k} \times \mathbf{A}_0 e^{i\psi}\} \tag{5.18}$$

Now, by taking the curl of Eq. 5.7, using Eq. 5.9 and then introducing Eq. 5.17 the following wave equation for A is obtained

$$\nabla^2 \mathbf{A} - \frac{1}{c^2}\frac{\partial^2}{\partial t^2}\mathbf{A} = -\frac{4\pi}{c}\mathbf{j}_{rot} \tag{5.19}$$

By now introducing (5.16) for $\mathbf{j}_{rot}$ the well known dispersion relation for electromagnetic waves in cold, collisionless plasmas is obtained:

$$\omega^2 = \omega_p^2 + k^2 c^2 \tag{5.20}$$

with the main result that there is a minimum frequency $\omega = \omega_p$ below which electromagnetic waves cannot propagate in the plasma. For a laser (vacuum) wavelength of 800 nm, the so-called critical density where the plasma frequency equals the laser frequency is reached at an electron number density of $1.74 \times 10^{21}/\text{cm}^{-3}$. With the definitions of the phase velocity $v_{ph} = \omega/k = c/\eta$ and the group velocity $v_g = \mathrm{d}\omega/\mathrm{d}k = c\eta$, the following expression for the refractive index η of the plasma can be obtained:

$$\eta = \sqrt{1 - \frac{\omega_p^2}{\omega^2}} \tag{5.21}$$

Therefore, a plasma has a refractive index smaller than one and its deviation from one grows with decreasing frequency of the incident wave. This leads to phase velocities larger than c and group velocities smaller than c. By using the ion mass and charge in Eq. 5.15 instead of electron mass and charge, the corresponding ion–plasma frequency ω_{ION} is obtained. It has been stated above that the ions are treated as immobile. This approximation is valid as long as the largest relevant timescale of the process under consideration is shorter than $1/\omega_{\text{ION}}$.

In order to derive the dispersion relation for electrostatic (charge separation) waves, one departs from the Poisson Eq. 5.11 together with the equation of motion of an electron fluid element and the continuity equation:

$$\nabla \cdot \mathbf{E} = -4\pi e(N - N_0) \tag{5.22}$$

$$m\frac{d\mathbf{u}}{dt} = -e\mathbf{E} \tag{5.23}$$

$$\frac{\partial N}{\partial t} + \nabla \cdot (N\mathbf{u}) = 0 \tag{5.24}$$

By looking for solutions for small deviations from quasi neutrality $\delta n = N - N_0 \ll N_0$ and linearizing and restricting to a one-dimensional description by setting

$$N(x,t) = N_0 + N_1(x,t) \tag{5.25}$$

$$\mathbf{u} = u(x,t)\hat{e}_x \tag{5.26}$$

$$\mathbf{E} = E_1(x,t)\hat{e}_x \tag{5.27}$$

and now searching for plane wave solutions by rewriting $N_1(x,t), u_1(x,t), E_1(x,t)$ in the form

$$\xi(x,t) = \xi e^{i(kx-\omega t)}, \quad \xi = N_1(x,t), u_1(x,t), E_1(x,t) \tag{5.28}$$

one finds the following system of algebraic equations

$$ikE_1 = -4\pi e N_1 \tag{5.29}$$

$$-i\omega m u_1 = -eE_1 \tag{5.30}$$

$$-i\omega N_1 + ikN_0 u_1 = 0 \tag{5.31}$$

which has non-trivial solutions for

$$\omega^2 = \frac{4\pi e^2 N_0}{m} = \omega_p^2 \tag{5.32}$$

Therefore, the oscillations take place at a constant frequency equal to the plasma frequency ω_p for all k vectors. For a warm plasma with a temperature T, this result is changed a bit with the dispersion relation reading (Meyer-ter-Vehn in [1])

$$\omega^2 = \omega_p^2 + 3k^2 v_{th}^2 \tag{5.33}$$

where v_{th} is the thermal electron velocity.

5.1.2 Relativistic Threshold Intensity

Usually, the range of validity of relativistic laser–plasma physics is given in terms of a threshold intensity beyond which relativistic effects start to play a significant role. This threshold is usually obtained in the following way from the non-relativistic equation of motion of a single electron:

$$m\frac{d\mathbf{u}}{dt} = -e\left(\mathbf{E} + \frac{\mathbf{u}}{c} \times \mathbf{B}\right) \tag{5.34}$$

Here $\mathbf{u}$ is the velocity of the single electron. $\mathbf{E}$ and $\mathbf{B}$ can again be expressed using the vector potential $\mathbf{A}$ as in Eqs. 5.5 and 5.6. By introducing now Eqs. 5.17 and 5.18 into Eq. 5.34 and after neglecting the term $\frac{\mathbf{u}}{c}\mathbf{B}$ against $\mathbf{E}$ which is justified in the non-relativistic case, one obtains after one trivial integration the following expression for the velocity of the electron:

$$\mathbf{u} = Re\left\{\frac{e\mathbf{E}}{im\omega}\right\} = -\frac{eA_0}{mc}\hat{e}\cos(\psi) \equiv a_0 c\cos(\psi). \tag{5.35}$$

Since (in this non-relativistic) calculation $u = |\boldsymbol{u}|$ is equal to c for $a_0 = eA_0/mc^2 = 1$, this is termed the relativistic threshold. Via $I_0 = \langle I \rangle = \langle|\mathbf{S}|\rangle = |(4\pi/c)\mathbf{E} \times \mathbf{B}| = \omega k/(8\pi)A_0^2$ this translates to a relativistic threshold Intensity I_0 of

$$I_0\lambda^2 = \frac{\pi}{2}P_0 a_0^2\,, \quad P_0 = \underbrace{\frac{mc^2}{e}}_{511\text{keV}} \cdot \underbrace{\frac{mc^3}{e}}_{17\text{kA}} \tag{5.36}$$

where 511 keV corresponds to the electron mass and 17 kA is the Alven current, the highest current that can be transported in vacuum without beam breakup caused by self-fields. P_0 is the relativistic power unit of 8.7 GW. In practical units, Eq. 5.36 reads

$$I_0\left(\frac{\text{W}}{\text{cm}^2}\right) \times \lambda^2(\mu\text{m}^2) = 1.37 \times 10^{18} \times a_0^2 \tag{5.37}$$

For a wavelength of $0.8\,\mu$m this gives a relativistic threshold intensity of $2.1 \times 10^{18}\,\text{W/cm}^2$.

5.1.3 Relativistic Single Electron in EM Field

The starting point for the analysis of relativistic motion of an electron in the electromagnetic field is the single electron Lagrangian [2]

$$L(\mathbf{r}, \mathbf{u}, t) = -mc^2\sqrt{1 - \beta^2} - \frac{e}{c}\mathbf{u} \cdot \mathbf{A} + \underbrace{e\Phi}_{=0} \tag{5.38}$$

with $\beta = \frac{u}{c}$. As above, the electromagnetic field is described by its vector potential $\mathbf{A}$, the scalar potential Φ being zero in the purely rotational electromagnetic wave of interest here. The electric and magnetic fields are connected to the vector potential again by Eqs. 5.17 and 5.18. The canonical (e.g. Lorentz invariant) momentum thus is

$$\mathbf{p}_{can} = \frac{\partial L}{\partial \mathbf{u}} = \frac{m\mathbf{u}}{\sqrt{1-\beta^2}} - \frac{e}{c}\mathbf{A} = \mathbf{p} - \frac{e}{c}\mathbf{A} \tag{5.39}$$

with the relativistic momentum $\mathbf{p} = m\gamma\mathbf{u}$ and $\gamma = 1/\sqrt{1-\beta^2} = \sqrt{1+(p/mc)^2}$. The Euler–Lagrange equations

$$\frac{d}{dt}\frac{\partial L}{\partial \mathbf{u}} - \frac{\partial L}{\partial \mathbf{r}} = 0 \tag{5.40}$$

now yield the equation of motion of a single electron in the electromagnetic field

$$\frac{d\mathbf{p}}{dt} = -e\left(\mathbf{E} + \frac{\mathbf{u}}{c} \times \mathbf{B}\right) \tag{5.41}$$

The energy of the electron $\mathcal{E}$ is given by its Hamiltonian $H = \mathbf{u} \cdot \mathbf{p}_{can} - L(\mathbf{r}, \mathbf{u}, t)$ thus being

$$\mathcal{E} = H = mc^2\gamma = mc^2\sqrt{1+\left(\frac{p}{mc}\right)^2} \tag{5.42}$$

Since $dH/dt = -\partial L/\partial t$ holds, H and, hence, $\mathcal{E}$ is constant if L has no explicit time dependence.

For a plane light wave, two symmetries exist providing two constants of motion (see Meyer-ter-Vehn et. al in [1]). Assuming plane waves, $\mathbf{A}(\mathbf{r}, t) \equiv \mathbf{A}(x, t)$ implies $\partial L/\partial r_\perp = 0$ and, thus, the transverse canonical momentum $\mathbf{p}_\perp$ is a constant of motion:

$$\mathbf{p}_{\perp,can} - \frac{e}{c}\mathbf{A}_{\perp,can} = \mathbf{p}_{\perp 0,can} \tag{5.43}$$

with $\mathbf{p}_{\perp 0,can}$ being the initial transverse momentum of the electron. For propagating waves $\mathbf{A}(\mathbf{r}, t) \equiv \mathbf{A}(t - x/c)$, the relation $dH/dt = -\partial L/\partial t$ yields

$$\mathcal{E} - p_{x,can}c = const. = mc^2 \tag{5.44}$$

as a constant of motion. With the kinetic energy being $\mathcal{E}_{kin} = mc^2(\gamma - 1) = p_{x,can}c = p^2_{\perp,can}/(2m)$ this gives the interesting result that the angle of propagation θ of the electron with respect to the propagation axis of the electromagnetic wave is a function of the electron gamma factor only and is given by

$$\tan\theta = \frac{p_{\perp,can}}{p_{x,can}} = \sqrt{\frac{2}{\gamma - 1}} \tag{5.45}$$

This has been experimentally confirmed in [3]. With the normalized vector potential $\mathbf{a} = e\mathbf{A}/mc^2$ it follows from (5.43) to (5.44) that the electron is accelerated to

$$\gamma = 1 + a^2/2 \tag{5.46}$$

if it was initially at rest and to

$$\gamma = \gamma_0 + a^2\gamma_0 \tag{5.47}$$

if it was already at relativistic speed corresponding to γ_0 before the interaction with the propagating wave. Therefore, in the latter case, the energy gain of the electron in the presence of the plane electromagnetic wave is much larger. It is emphasized here that these results do not correspond to a net energy gain by the electron but rather represent the instantaneous dependence of the electron motion on the electromagnetic wave. Once the plane wave has passed, the electron is at rest again if initially at rest. It is only displaced along the propagation axis of the wave a certain distance. However, *if* the electron did experience a net energy transfer, thereby obtaining a certain kinetic energy, and still the plane propagating wave approximation is—at least approximately fulfilled—then it will propagate at the angle θ, as prescribed by Eq. 5.45. The question under which circumstances a single, free electron in vacuum can actually be accelerated by an electromagnetic wave was treated by Woodward [4] and Lawson [5]. They showed that the net energy gain of a relativistic electron interacting with an electromagnetic field in vacuum is zero if the following assumptions are made [6]:

1. The laser field is in vacuum and is not truncated by any boundaries.
2. The electron is highly relativistic along the acceleration path.
3. There are no static electric or magnetic fields.
4. The interaction region is infinite.
5. Ponderomotive effects are neglected.

As soon as one or more of these criteria are broken, electrons can in principle gain energy from the laser field. An account of several methods that allow to accelerate electrons in vacuum is given in [6]. For example, in [7] it is shown that under certain conditions strong focusing is enough to obtain net acceleration. So far, however, vacuum acceleration schemes have failed to show significant acceleration of sizeable amounts of electrons.

In contrast, a very efficient way of accelerating large numbers of electrons with relativistically strong transversal electromagnetic waves works via the generation of strongly driven Langmuir waves in a plasma, a situation which certainly violates all of the assumptions made in the Lawson–Woodward theorem. This mechanism is outlined in the following sections.

5.1.4 Relativistic Cold Collisionless Plasma Equations

The *cold plasma* equations given in the following describe the plasma as an electron fluid with density $N(\mathbf{x}, t)$ with zero temperature that moves on a background of immobile neutralizing ions with a density of $N_i = N_0$. All plasma parameters are in Eulerian coordinates, thus, all parameters are functions of position **r** and time t. The following dimensionless variables are introduced for convenience

$$a = \frac{e\mathbf{A}}{mc^2} \quad \varphi = \frac{e\Phi}{mc^2} \quad \mathbf{p} = \frac{\mathbf{P}}{mc} \quad n = \frac{N}{N_0} \tag{5.48}$$

Here $\mathbf{P} = m\gamma\mathbf{u}$. The current density is given by $\mathbf{J} = -eN\mathbf{v} = -eN\mathbf{P}/m\gamma$, with $\gamma = \sqrt{1+p^2}$.

The equation of motion of the electron fluid is

$$\frac{d}{dt}\mathbf{P}(\mathbf{r}, t) = -e\left(\mathbf{E} + \frac{\mathbf{u}}{c} \times \mathbf{B}\right) \tag{5.49}$$

where, in the fluid description, d/dt is the convective differential in Eulerian description

$$\frac{d}{dt} = \left(\frac{\partial}{\partial t} + \mathbf{u} \cdot \nabla\right) \tag{5.50}$$

that essentially states that a certain parameter of a fluid element not only changes due to an explicit dependence on time but also because fluid from neighboring positions is flowing in. Using this and the potentials **A** and **Φ** introduced in Sect. 5.1.1 Eq. 5.49 becomes

$$\left(\frac{\partial}{\partial t} + \mathbf{u} \cdot \nabla\right)\mathbf{P} = -e\left(-\frac{1}{c}\frac{\partial \mathbf{A}}{\partial t} - \nabla\Phi + \frac{\mathbf{u}}{c} \times (\nabla \times \mathbf{A})\right) \tag{5.51}$$

By introducing now the dimensionless variables (5.48) and noting that $(\mathbf{u}/c \cdot \nabla)\mathbf{p} = \nabla\gamma - \mathbf{u}/c \times (\nabla \times \mathbf{p})$ and $\nabla\gamma = \nabla\sqrt{1+p^2} = 1/(2\gamma)\nabla p^2$, the relativistic equation of motion of cold collisionless plasmas is obtained:

$$\frac{1}{c}\frac{\partial}{\partial t}(\mathbf{p} - \mathbf{a}) - \frac{\mathbf{u}}{c} \times [\nabla \times (\mathbf{p} - \mathbf{a})] = \nabla(\varphi - \gamma) \tag{5.52}$$

In the adiabatic approximation $\mathbf{p} \simeq \mathbf{a}$ one basic solution is obtained as $\nabla(\varphi - \gamma) = 0$, stating that the electrostatic force $\nabla\varphi$ is balanced exactly by the relativistic ponderomotive force $\propto \nabla\gamma$ which in the adiabatic case is given as

$$F_{pond} = -Nmc^2\nabla\gamma = -Nmc^2\frac{1}{\gamma}\nabla\frac{|a|^2}{2} \tag{5.53}$$

The proof that the above expression corresponds to the relativistic ponderomotive force is non-trivial [8, 9] and it also produces some counterintuitive effects like, for example, uphill acceleration [8]. This difficulties encountered here do not come surprising taking into account that the classical ponderomotive force is a residual effect found by first order perturbation theory applied to the Lorentz force and, thus, requiring *small* particle velocities and correspondingly *small* electromagnetic fields.

Going back to the non-normalized quantities in Eq. 5.53 and using a harmonic plane wave as in 5.17 the following expression for the relativistic ponderomotive force is obtained

$$F_{pond} = -\frac{\omega_p^2}{\omega^2\gamma}\nabla\frac{|E_L|^2}{8\pi} \tag{5.54}$$

In the limit $\gamma \to 1$, the classic result is recovered. Here ω and E_L are frequency and electric field strength of the harmonic electromagnetic wave. If the relativistic orbit of a particle is to be calculated using the relativistic ponderomotive force, of course the relativistic formalism must be applied and, furthermore, it must be taken into account that the ponderomotive force is attached to the (relativistic) movement of the average center of gravity of a rapidly oscillating particle. This can elegantly be accomplished by ascribing an effective mass to this center of gravity. Details and examples can be found in [8].

In order to obtain a complete description of the cold relativistic plasma, besides (5.52) two more relations are needed, namely the Poisson equation

$$\nabla^2\varphi = \frac{\omega_p^2}{c^2}(n-1) \tag{5.55}$$

and, as always in fluid dynamics, the continuity equation

$$\frac{\partial n}{\partial t} + \nabla(n\mathbf{u}) = 0 \tag{5.56}$$

using these relations, from ampere's law (5.3) and using Coulomb Gauge ($\nabla \cdot \mathbf{A} = 0$) the following wave equation is derived:

$$\left(\nabla^2 - \frac{1}{c^2}\frac{\partial^2}{\partial t^2}\right)\boldsymbol{a} = \frac{\omega_p^2}{c^2}\cdot\frac{n\boldsymbol{p}}{\gamma} + \frac{1}{c}\frac{\partial}{\partial t}\nabla\varphi \tag{5.57}$$

Only two special solutions of this equation, which are of primary interest in the context of laser driven electron acceleration, are considered in the following. These are propagating electromagnetic plane waves, studied in Sect. 5.1.5 and electrostatic waves, analyzed in Sect. 5.1.6. A comprehensive compilation of analytic results on travelling waves in relativistic cold collisionless plasmas is given in [10].

5.1.5 Electromagnetic Waves: Self-Focusing

In order to treat electromagnetic waves, the electric field is separated into a rotational part and gradient part as in Eq. 5.5. Then from (5.57) the following wave equation for electromagnetic waves is obtained

$$\left(\nabla^2 - \frac{1}{c^2}\frac{\partial^2}{\partial t^2}\right)\boldsymbol{a} = \frac{\omega_p^2}{c^2}\cdot\frac{n\boldsymbol{p}}{\gamma} \tag{5.58}$$

The analysis is restricted now to transverse electromagnetic pulses with small and slowly varying amplitude a of the form

$$\boldsymbol{a} = \frac{a}{2}\boldsymbol{\sigma} e^{i\psi} + c.c., \quad \psi = i(\boldsymbol{k}r - \omega t) \tag{5.59}$$

where $c.c.$ signifies the conjugated complex and where the polarization vector $\boldsymbol{\sigma}$ is defined as $\boldsymbol{\sigma} = \boldsymbol{e}_y$ for linear polarization and $1/\sqrt{2}(\boldsymbol{e}_y + i\boldsymbol{e}_y)$ for circular polarization. The vectors $\boldsymbol{e}_x, \boldsymbol{e}_y, \boldsymbol{e}_z$ are unit vectors in x, y, and z direction, respectively. With $\boldsymbol{\sigma}^2 = 0$ and $\boldsymbol{\sigma\sigma}^* = 1$, the square of the pump amplitude is now given by

$$\boldsymbol{a}^2 = \begin{cases} \frac{1}{4}a^2 e^{2i\psi} + \frac{1}{2}|a|^2 + \frac{1}{4}a^{*2}e{-}i2\psi, & \text{linear} \\ \frac{1}{2}|a|^2, & \text{circular} \end{cases} \tag{5.60}$$

The non-vanishing rapidly oscillating terms in the case of linear polarization complicate the analysis, therefore, in the following only circular polarization is treated. The Poisson Eq. 5.55 in adiabatic approximation [see Eq. (5.52)]

$$n = 1 + \frac{c^2}{\omega_p^2}\nabla^2\varphi \simeq 1 + \frac{c^2}{\omega_p^2}\nabla^2\gamma \tag{5.61}$$

shows, that the (normalized) density n only depends in higher orders on γ, therefore, a first order result can be obtained from Eq. 5.58 by setting $n = 1$ and using the adiabatic approximation $\boldsymbol{a} = \boldsymbol{p}$. Then, by introducing 5.59, the relativistic dispersion relation is obtained

$$\omega_0^2 = c^2 k_0^2 + \omega_p^2/\gamma \tag{5.62}$$

containing now a relativistically modified plasma frequency $\omega_p/\sqrt{\gamma}$. This modification leads to the so-called relativistically induced transparency: laser pulses of sufficient intensity are able to propagate in plasmas with over-critical plasma density because the plasma frequency is reduced by a factor of $\sqrt{\gamma}$. By substituting now ω_p^2/γ for ω_p^2 in Eq. 5.21, the relativistically modified refractive index reads as

$$\eta(r) = \sqrt{1 - \frac{\omega_p}{\omega\sqrt{1 + a(r)^2/2}}} \tag{5.63}$$

Here, the gamma-factor of the electron fluid in the adiabatic limit and with the electromagnetic wave having circular polarization has been used:

$$\gamma = \sqrt{1 + \boldsymbol{a}^2} = \sqrt{1 + \frac{a^2}{2}} \tag{5.64}$$

Equation 5.63 states that the radial intensity variation of the electromagnetic wave leads to a radially changing refractive index which will start to counteract the natural diffraction of the beam. In order to get a qualitative picture of this self-focusing, the diffraction of a gaussian beam can be balanced against the self-focusing effect in the following way [11]: the envelope of a Gaussian beam with a peak normalized vector potential of a_0 in the focus is given by [12]

$$a(r,z) = \frac{a_0 e^{-r^2/\left(\sigma_0^2\left(1+z^2/Z_R^2\right)\right)}}{\sqrt{1+\frac{z^2}{Z_R^2}}} \exp\left\{ -i \arctan\left(\frac{z}{Z_R}\right) + i\left(\frac{r}{\sigma_0}\right)^2 \frac{z/Z_R}{1+z^2/Z_R^2} \right\} \quad (5.65)$$

with σ_0 being the focal spot radius and the Rayleigh range Z_R being the distance from focus at which the beam cross section is double that of the focal spot cross section

$$Z_R = k\frac{\sigma_0^2}{2} = \frac{\pi\sigma_0^2}{\lambda} \quad (5.66)$$

Then, the beam diffracts from the focal spot with the divergence angle θ

$$\frac{dR}{dZ} = \frac{\sigma_0}{Z_R} = \frac{1}{k\sigma_0} \quad (5.67)$$

Counteracting this diffraction, the dependence of the refractive index on the laser intensity 5.63 causes the refractive index to be larger on-axis than off-axis thereby creating a lensing effect. To estimate the magnitude of this focusing effect, one may compare the phase velocity on-axis where the intensity is highest and off-axis. The larger off-axis phase velocity will then bend the phase fronts and lead to beam contraction. The radius dependent phase velocity $v_P(r)$ is given by

$$\frac{v_P(r)}{c} = \frac{1}{\eta(r)} \simeq 1 + \frac{\omega_P^2}{2\omega^2}\left(1 - \frac{a^2(r)}{4}\right) \quad (5.68)$$

such that the velocity difference between the beam axis and at a radial distance r from the axis reads

$$\frac{\Delta v_P(r)}{c} = \frac{\omega_P^2}{8\omega^2} a_0^2 e^{-r^2/\sigma_0^2} \quad (5.69)$$

The maximum path difference is then given by

$$\Delta L = |\Delta v_P|_{max}\, t = |\frac{\Delta v_P}{c}|_{max}\, Z = \alpha R, \quad \alpha^2 = \frac{\omega_P^2 a_0^2}{8\omega^2} \quad (5.70)$$

where R is the initial beam radius. By requiring now that $\theta \equiv \alpha$ self-focusing just cancels diffraction. This results in a threshold value for the laser beam power $P_L \propto a_0^2\sigma_0^2$:

$$a_0^2\left(\frac{\omega_P\sigma_0}{c}\right)^2 \geq 8 \tag{5.71}$$

Rigorous derivations of the the critical power $\mathcal{P}_C$ where relativistic self-focusing and beam diffraction are just balanced can be found in [13–15], the value usually stated in the literature is

$$\mathcal{P}_C = 2.0\left(\frac{\omega}{\omega_p}\right)^2 \mathcal{P}_0, \quad \mathcal{P}_0 = \frac{m^2c^5}{e^2} = 8.7\ \text{GW} \tag{5.72}$$

By further increasing the power of the light pulse, electrons are pushed out by the ponderomotive force, readily leading to complete cavitation. This regime has been studied by numerically solving the differential equations [16–18], basically showing that under certain conditions an oscillatory behavior is observed and that for complete cavitation a finite channel—and laser pulse—diameter of $D \approx 4(c/\omega_p)$ is obtained. Later, when computing resources allowed full 3D particle-in-cell (PIC) simulations, it was confirmed that self-focusing in 2D and 3D show very different behavior [19] and it was shown that complete cavitation is hindered by strong electron heating in the channel and strong light scattering before the channel is formed [20, 21].

The intensity dependence of the refractive index in the relativistic intensity regime $a \geq 1$ leads also to a variation of the refractive index and hence the phase and group velocity in longitudinal direction, causing self phase modulation of the pulse. A similar effect is caused by the electron density variation at the leading edge of the pulse, which also causes a varying index of refraction in longitudinal direction. Via the concept of photon acceleration this can directly be linked to the energy transfer process by which laser energy is transferred to the wake causing a spectral redshift of the laser light [22].

5.1.6 Electrostatic Waves: Wave Breaking

The following description follows [23]. The electrostatic equation of motion for a cold relativistic plasma is given by

$$\left(\frac{\partial}{\partial t} + \boldsymbol{u}\cdot\nabla\right)(\gamma m\boldsymbol{u}) = -e\boldsymbol{E} \tag{5.73}$$

which together with the continuity Eq. 5.56 and the Poisson equation

$$\nabla\cdot\boldsymbol{E} = 4\pi e(N_0 - N) \tag{5.74}$$

form the basis of the following derivation. Here, N_0 is the constant background density of the ions. Restricting to a one dimensional analysis and only considering propagating waves which depend on x and t only through τ, with $\tau = \omega_p(t - x/v_{ph})$

and where v_{ph} is the phase velocity of the electrostatic plasma wave, the following set of equations is obtained

$$\frac{d}{d\tau}N(\tau) = \frac{1}{v_{ph}}\frac{d}{d\tau}(N(\tau)u(\tau)) \tag{5.75}$$

$$\frac{d}{d\tau}(\gamma(\tau)mu(\tau)) = -\frac{eE}{\omega_p\left(1 - \frac{u(\tau)}{v_{ph}}\right)} \tag{5.76}$$

$$\frac{d}{d\tau}E(\tau) = -\frac{4\pi e v_{ph}}{\omega_p}(N_0 - N(\tau)) \tag{5.77}$$

From now on, the argument τ is suppressed for clarity. From these equations one gets by introducing the normalized density $n = N/N_0$ and normalized field $\hat{E} = eE/m\omega_p c$, and with $\beta_{ph} = v_{ph}/c$ and $\beta = u/c$:

$$n = \frac{1}{1 - \frac{\beta}{\beta_{ph}}} \tag{5.78}$$

$$-\hat{E} = \left(1 - \frac{\beta}{\beta_{ph}}\right)\frac{d}{d\tau}(\gamma\beta) \tag{5.79}$$

$$\frac{d}{d\tau}\hat{E} = \frac{\beta}{1 - \frac{\beta}{\beta_{ph}}} \tag{5.80}$$

Equation 5.78 shows a central feature of longitudinal plasma waves: when the electron fluid velocity β approaches the phase velocity of the wave β_{ph}, the density diverges. This corresponds to the breaking of the wave, an effect that cannot be treated within the fluid approximation because it leads to multi-flow conditions (at the same spatial coordinate, fluid packets move at different velocities) that must be treated within a kinetic description—either by solving the Vlasov-equation or by PIC codes. The greater β_{ph} is, the harder it becomes to get wave breaking, as the electron-beta cannot exceed unity. In the context of laser wakefield acceleration, where β_{ph} is always close to one this is a beneficial effect because it renders the resultant wake wave very robust.

By combining the Euler and Poisson equations (5.79) and (5.80), the first integral is obtained as

$$\hat{E}(\gamma) = \sqrt{2(\gamma_m - \gamma)} \tag{5.81}$$

where the integration constant has been set to $\gamma_m = 1/\sqrt{1 - \beta_m^2}$ corresponding to the maximum fluid velocity. For $\gamma = \gamma_m$ the electric field vanishes, and for $v = 0$ one has $\gamma = 1$ and the maximum field is obtained as

$$\hat{E}_{max} = \sqrt{2(\gamma_m - 1)} \tag{5.82}$$

Using the following relation for γ_{ph}

$$\gamma_{ph} = \frac{1}{\sqrt{1 - \frac{v_{ph}^2}{c^2}}} = \frac{1}{\sqrt{1 - \frac{v_G^2}{c^2}}} = \frac{\omega}{\omega_P} = \sqrt{\frac{N_C}{N_0}} \tag{5.83}$$

with v_G being the group velocity of the laser pulse given by

$$v_G = c\eta = c\sqrt{1 - \frac{\omega_p^2}{\omega_L^2}} \tag{5.84}$$

in the relativistic case $\gamma \gg 1$ one gets for a phase velocity that equals the group velocity of the laser pulse a maximum field E_{WB} for which wave breaking sets in of

$$E_{WB} = \sqrt{2(\gamma_{ph} + 1)E_0} \cong \sqrt{2\gamma_{ph}\frac{m\omega_p c}{e}} = \sqrt{2}2\pi\frac{mc^2}{e\lambda_L}\left(\frac{\omega_P}{\omega_L}\right)^{\frac{1}{2}} \approx 5 \times 10^{12}\frac{V}{m}\left(\frac{N_0}{N_C}\right)^{\frac{1}{4}} \tag{5.85}$$

Here, the approximate numerical value for the wave breaking limit is given for typical laser wavelengths between 800 and 1,000 nm.

As shown in [23], also a warm relativistic plasma can be treated analytically, the wave breaking limit in this case is for $v_{ph} = c$

$$E_{WB} \approx \frac{m\omega_p c}{e}\left(\frac{4}{27}\frac{mc^2}{T}\right)^{\frac{1}{4}} \tag{5.86}$$

This shows that for rising temperature the wave breaking threshold drops. This was to be expected because in a warm plasma there is always a small population of electrons fast enough to get trapped and this population grows of course with the temperature.

Electrons which are externally injected into the plasma wave or self-injected due a wave-breaking process (see the following section), are accelerated by the longitudinal electric wake field. Due to the very large fields in the range of GV/m–TV/m, the highly relativistic regime, where the electron velocity is very close to c is readily reached. The phase velocity of the wake wave, however, equals the group velocity of the laser pulse, and this group velocity in turn is smaller than c according to $v_g = c\eta$, with η being the refractive index of the plasma (5.21). This small deviation of v_{ph} from c has been neglected in the derivation above. Therefore, after a propagation distance corresponding to the so-called *dephasing length* L_d, the electrons outrun the wake and start to be decelerated again giving their energy back to the wake wave. The dephasing length is then given by the length needed by the injected electron to gain half a plasma wavelength with respect to the wake wave:

$$L_d = \frac{\pi}{\omega_p(1/v_{ph} - 1/c)} = \frac{\pi}{\frac{2\pi}{\lambda_p}(\frac{1}{\beta_{ph}} - 1)} \approx \lambda_p \gamma_{ph}^2 \quad (5.87)$$

This corresponds to a maximum energy $\mathcal{E}_{max}$ of the accelerated electrons at the cold relativistic wave breaking limit of

$$\mathcal{E}_{max} = E_{WB} \lambda_p \gamma_{ph}^2 \approx 5 \text{ MeV} \times \left(\frac{\omega}{\omega_p}\right)^{5/2} \quad (5.88)$$

5.1.7 Laser Wakefield Acceleration and Scaling Laws

In the last section, one-dimensional wave breaking was summarized. It was shown, however, that in general the wave breaking thresholds are not the same in 1D, 2D and 3D [19] owing to transversal wave breaking (see Sect. 5.2) that sets in long before the longitudinal one. This is actually a beneficial effect in current laser plasma accelerators because it facilitates self-injection of electrons into laser-driven plasma waves far below the cold longitudinal wave breaking limit. While only this has allowed the breakthrough results in laser driven electron acceleration [24–26] it brings with it the disadvantage that there is no way of controlling the number of trapped electrons or the positions where trapping starts and stops (see Chap. 8).

Since the kinetic nature (multi-stream flows) and the complexity of three dimensional relativistic laser driven plasma waves has so far defeated rigorous analytical treatment, prediction of the properties of these systems relies strongly on numeric simulation by particle-in-cell(PIC) codes. These codes describe the system by simulating the motion of quasi-particles, each of them representing an electron cloud containing thousands of electrons. In order to keep the computational requirements in a realistic regime, no particle-to-particle interactions are considered because this would lead to a scaling of the computational cost of the simulation with N^2 where N is the number of macro-particles. Instead, the charge and current densities caused by the macro-particles are calculated and interpolated onto a grid that fills the simulation space. Then, the electric and magnetic fields generated by these charges and currents are calculated and stored at the nodes of the grid as well. Afterwards the macro-particles are propagated by interpolating the fields of the close-by cell nodes to the current position of the particle and, using a short enough time step, the new position is computed. Now, the charge and current densities are calculated again, and the cycle starts anew. In this way, the computationally prohibitively expensive tasks of considering the particle-to-particle interactions, or the one of numerically solving the six-dimensional Vlasov equation, is circumvented.

Numerical studies using such PIC-codes [27–29] have predicted for a certain parameter regime the emergence of monoenergetic electrons bunches with

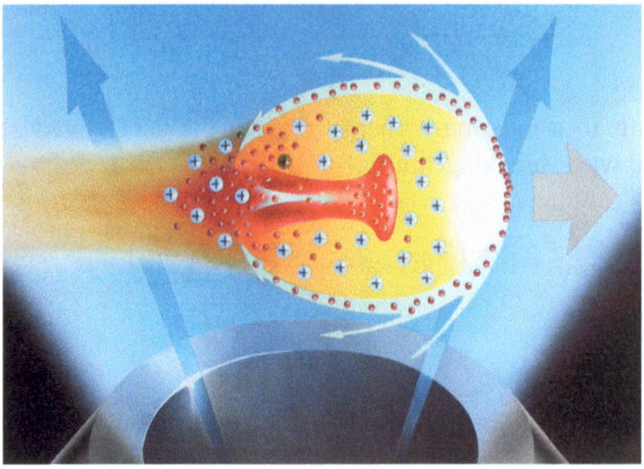

Fig. 5.1 Artist's conception of electron acceleration in the bubble-Regime.The laser pulse, shown in *white*, propagates from left to right and, almost like a *light-bullet*, pushes the electrons (*red*) transversally out of its path. The heavy ions (*blue*) are left behind, their strong electric field pulls some of the electrons into the bubble transversally and then accelerates them to ultra-relativistic energies

relativistic energies and few-femtosecond duration from laser excited plasma waves. In this regime, the plasma waves are strongly driven by a laser pulse of ultra-relativistic intensity ($\gtrsim 10^{19}\,\mathrm{W/cm^2}$) and a duration and diameter that are matched to the plasma density. This means that ideally a laser pulse is required that has a length of equal to or less than half the plasma wavelength $\lambda_p/2 = \pi c/\omega_p$ and a (focal) diameter of one λ_p. An artist's conception of this acceleration scheme is shown in Fig. 5.1. The laser pulse, shown in white, is so intense that its ponderomotive force pushes free electrons (red) transversally out of its path leaving the positively charged and much heavier ions (blue) behind. The electrons are pulled back to the axis by the electrical field created by charge separation after a propagation length comparable to a plasma wavelength. In that way, they form a cavity void of electrons trailing the laser pulse, dubbed a "bubble" [27], which constitutes a highly anharmonic Langmuir wave that breaks down completely after the first period of its oscillation. According to the simulations, a fraction of the returning electrons are trapped in the bubble and accelerated by its strong longitudinal electric field, producing relativistic electron bunches with narrow-band energy spectra [27, 30, 31].

Gordienko and Pukhov also developed an analytical similarity theory for ultra-relativistic ($a_0 > 1$) laser–plasmas which will be shortly presented in the following because it gives predictions for the parameter regime of interest in the present experiments.

This theory is conceptually equivalent to well known similarity laws in hydro- and gas dynamics. Similarly to gas dynamical scalings presented in part I of this work, with the dimensionless Knudsen number *Kn* as a similarity parameter, here, a general scaling with a dimensionless similarity parameter *S* was found that determines the characteristics of the laser plasma interaction. This similarity parameter is given by

$$S = \frac{N}{a_0 N_c} \tag{5.89}$$

with all the symbols defined as in the last section. Strongly underdense plasmas ($N/N_c \ll 1$) as they occur in laser wake field acceleration hence correspond to the limit $S \ll 1$. In this limit, Gordienko and Pukhov derived scaling laws obtaining the parametric dependencies from their similarity theory and the numerical constants and pre-factors from comparison with a number of PIC-simulations. These scaling laws are valid for the bubble regime and are presented in [30] and [31], they are summarized here:

The optimal radius w_0 (intensity $1/e^2$—radius) of the laser pulse must be matched to the bubble radius R:

$$k_0 R \approx 1/\sqrt{S} \rightarrow R \approx \sqrt{\frac{mc^2 a_0}{4\pi N e^2}} \tag{5.90}$$

This is related to the plasma density via

$$k_p R \simeq k_p w_0 = \varsigma \sqrt{a_0} \tag{5.91}$$

a relation that essentially states equilibrium between ponderomotive force pushing the electrons out of and electrostatic force drawing the electrons towards the bubble [30]. ς is equal to one for circular polarization and equal to 1.12 for linear polarization [32]. Of course, a real world laser has constant pulse power and not constant intensity, therefore, it is useful to cast Eq. 5.91 in the following form

$$a_{0,opt} = \left(\frac{32}{\varsigma^2}\right)^{\frac{1}{3}} \left(\frac{\mathcal{P}}{\mathcal{P}_c}\right)^{\frac{1}{3}} \tag{5.92}$$

giving the optimal a_0 for a given pulse power P and density (through $\mathcal{P}_c(N_e)$). ς is a polarization dependent parameter, equal to 1.12 for linear polarization and equal to one for circular polarization. In order to obtain Eq. 5.92 from (5.91), formula (5.72) and the relation between beam power $\mathcal{P}$ and a_0 of a gaussian beam has to be used [1]:

$$\mathcal{P} = \frac{\omega^2}{16c^2} a_0^2 R^2 \mathcal{P}_0, \quad \mathcal{P}_0 = \frac{m^2 c^5}{e^2} \tag{5.93}$$

Now, with $a_{0,opt}$ given, through (5.92) the optimal focal spot radius $w_{0,opt}$ can be calculated, it evaluates to

$$w_{0,opt} = \frac{c}{\omega_p} \varsigma^{\frac{2}{3}} 32^{\frac{1}{6}} \left(\frac{\mathcal{P}}{\mathcal{P}_c}\right)^{\frac{1}{6}} \tag{5.94}$$

In practical units this gives for the intensity full width at half maximum diameter of the optimal focal spot $d_{0,opt}^{FWHM}$ and for linear polarization

$$d_{0,opt}^{FWHM} = 3.4 \times \left(\frac{\lambda}{N}\right)^{\frac{1}{3}} \times \mathcal{P}^{\frac{1}{6}} \tag{5.95}$$

where $[d_{0,opt}^{FWHM}] = [\lambda] = \mu\text{m}$, $[N] = 10^{19}/\text{cm}^3$ and $[\mathcal{P}] = TW$.

For efficient acceleration and mono-energetic spectra with small energy spread, the laser pulse duration (intensity $1/e^2$-duration) must satisfy

$$\tau c \leq R \tag{5.96}$$

The electron density N must lie between a lower and an upper bound, where the lower bound is set by the requirement that the laser pulse is ultra-relativistic $a_0 > 1$ and the upper limit is implicitly defined by the condition (5.90), yielding

$$N_c \frac{\mathcal{P}_0}{\mathcal{P}} < N < N_c \sqrt{\frac{\mathcal{P}}{\mathcal{P}_0}} \frac{1}{(\omega_L \tau)^3} \tag{5.97}$$

In order to reach the bubble regime, the laser pulse power $\mathcal{P}$ must be larger than a threshold value of

$$\mathcal{P} > \mathcal{P}_{Bubble} (\omega_L \tau)^2 = 30\text{GW} \times \left(\frac{\tau[\text{fs}]}{\lambda[\mu\text{m}]}\right)^2 \tag{5.98}$$

With all these requirements being fulfilled, a mono-energetic electron bunch is then produced, obeying the following scalings for the electron energy $\mathcal{E}_{mono}$, the number of electrons in the bunch $\mathcal{N}_{mono}$, the acceleration length L_{acc} and the energy transfer efficiency between laser pulse and electron bunch ζ:

$$\mathcal{E}_{mono} \approx 0.65 mc^2 \sqrt{\frac{\mathcal{P}}{\mathcal{P}_0}} \frac{c\tau}{\lambda} \tag{5.99}$$

$$\mathcal{N}_{mono} \approx \frac{1.8}{k_0 r_e} \sqrt{\frac{\mathcal{P}}{\mathcal{P}_0}} \tag{5.100}$$

$$L_{acc} \approx 0.7 \frac{c\tau}{\lambda} Z_R \tag{5.101}$$

$$\zeta = \frac{\mathcal{E}_{mono} \mathcal{N}_{mono}}{\tau \mathcal{P}} \approx 20\% \tag{5.102}$$

Here, $Z_R = \pi R^2/\lambda$ is the Rayleigh range of the focused Gaussian beam, $r_e = e^2/mc^2$ is the classical electron radius, and $k_0 = 2\pi/\lambda$. It is interesting to note that the efficiency is predicted to be constant $\approx 20\%$ for all cases within the bubble regime.

For laser pulses with a central wavelength of 800 nm, a pulse duration of 8 fs, and 40 mJ pulse energy, corresponding to 5 TW power the upper and lower limits for the electron density N, according to relation (5.97) give $3 \times 10^{18} < N < 6 \times 10^{18}\,\mathrm{cm}^{-3}$. However, the experimentally obtained optimal value is $2 \times 10^{19}\,\mathrm{cm}^{-3}$. For these values the theory predicts an optimum focal spot size of 3.3 μm (FWHM) corresponding to a peak intensity of $3.5 \times 10^{19}\,\mathrm{W/cm^2}$. The power threshold for the bubble regime is around 3.1 TW corresponding to a pulse energy of 25 mJ. The considered pulses have 5 TW and, therefore, produce a bubble that is predicted to yield electron bunches with an energy of 24 MeV and a charge of 312 pC. The acceleration length L_{acc} is then approximately $L_{acc} \approx 68\,\mu\mathrm{m}$. Comparison with the experimental results shows that the experimentally obtained energy is in the range of 20 MeV and hence confirms the validity of the energy scaling (5.99). The acceleration length can be estimated from the experiment to be within 50 to maximum 150 μm agreeing with the scaling (5.101). The experimentally determined charge of the accelerated electron bunch is between 1 and 10 pC. This is in clear contradiction to the prediction by formula (5.100), which gives a two orders of magnitude higher result. This also causes a large deviation of predicted efficiency of approximately 20% to the experimental value of below 1%. For details on the experimental results see Chap. 7.

For the sake of completeness, the electron beam parameters as predicted by the above scaling laws for the upgraded laser system with double the laser pulse energy are given here as well. These modified values are an optimum spot size of 3.7 μm, an optimum density range of $1.5 \times 10^{18} < N < 1.5 \times 10^{19}\,\mathrm{cm}^{-3}$ corresponding to a peak intensity of $2.9 \times 10^{19}\,\mathrm{W/cm^2}$, an electron energy of 34 MeV, 442 pC charge, and an acceleration length of 78 μm.

A different set of scaling laws based on an empirical theory is given by Lu et al. [28, 32]. It gives in general larger optimal focal spots and longer optimal pulse durations. For the above 5 TW laser, the optimal spot is here 11 μm FWHM and the predicted electron energy is around 80 MeV. These values are clearly far away from the experimentally obtained optimum values, so this model does not seem to be suitable for estimating the performance of current or future electron accelerators using few-cycle laser pulses. The reason for this discrepancy seems to rest in the central role that the "etching rate" has in the model by Lu et al. As the pulse propagates through the plasma, its leading edge constantly looses energy to the plasma electrons thereby "etching" backward into the pulse. The assumption by Lu et al. is now that acceleration goes on until the backward-etching pulse front arrives at the trailing pulse edge. The time that this process takes is simply calculated by dividing the etching speed by the pulse duration. While this seems to be a valid assumption for cigar-shaped pulses, it clearly is not applicable for the ultrashort pulses used in the present work. On the contrary, our PIC simulations show that the pulse duration is reduced by only 10% after the interaction and that the pulse simply stops being able to drive the bubble due to intensity-loss caused by depletion and (mainly) diffraction.

The conclusion here seems to be that each of the models is the most accurate in different cases: the model by Gordienko and Pukhov works best for pulses that are roughly half as long as their diameter and higher plasma densities, whereas the model by Lu et al. is accurate for pulses with longer pulse durations and lower densities.

5.2 Results of Particle-In-Cell Simulations

In this section, results of PIC simulations are presented that were performed by Michael Geissler with his three dimensional PIC code ILLUMINATION [33]. The co-moving simulation box represents a volume of $27 \times 27 \times 36\,\mu m$ (z is the propagation axis) with a grid size of $94 \times 94 \times 47\,nm$. The plasma is treated as fully (pre-)ionized with one macro-particle per cell and immobile ions as a charge neutralizing background. It is modelled with a uniform transversal density distribution whereas in the longitudinal direction a $120\,\mu m$ broad flat-top profile, matching the experimentally determined channel length, with an electron density of $2 \times 10^{19}\,cm^{-3}$ terminated by exponential gradients, are assumed. The $1/e$ scale length of the entrance gradient is $5\,\mu m$, just to avoid numerical problems with a steep gradient. The exit gradient was chosen as $30\,\mu m$ to match the experimental conditions. The simulations were performed with a time step of 0.1 fs using the following laser parameters: pulse duration: 8.5 fs (Intensity-FWHM), spot size (Intensity-FWHM): $7\,\mu m$, pulse energy: 38 mJ, and carrier laser wavelength: 800 nm. These closely match the typical experimental conditions.

Figure 5.2 shows snapshots of the electron density distribution, laser intensity, and longitudinal electrical field of the plasma accelerator at three different positions, namely after the laser pulse has travelled 23 (a, b), 57 (c, d) and 123 (e, f) μm inside the plasma. In panels (a) and (b), the laser pulse has just entered the plasma immediately causing strong charge separation and thereby exciting a plasma wave. The charge separation in turn causes a strong longitudinal electrostatic field to arise, its on-axis value is plotted in red. Already at this early stage in the laser propagation, the plasma wave is strongly anharmonic, having the typical saw-tooth shape of strongly driven Langmuir waves. In panel (b) it can be seen that the electron density in the first through of the wave is reduced to approximately 35% and also the imprint of the laser pulse on the plasma density is clearly visible. At the back of the first plasma oscillation, there is a density peak having four times the density of the background plasma. This is a typical feature of strongly driven plasma waves. It has been shown above that in the fluid model of the plasma, the density diverges just at the moment when the fluid velocity approaches the phase velocity of the plasma wave, leading to wave-breaking and of course also to the breakdown of the fluid model. The strong density peaks in the plasma wave as well as the saw-tooth shape of the longitudinal electric field therefore clearly indicate that the laser pulse is strong enough to drive the plasma

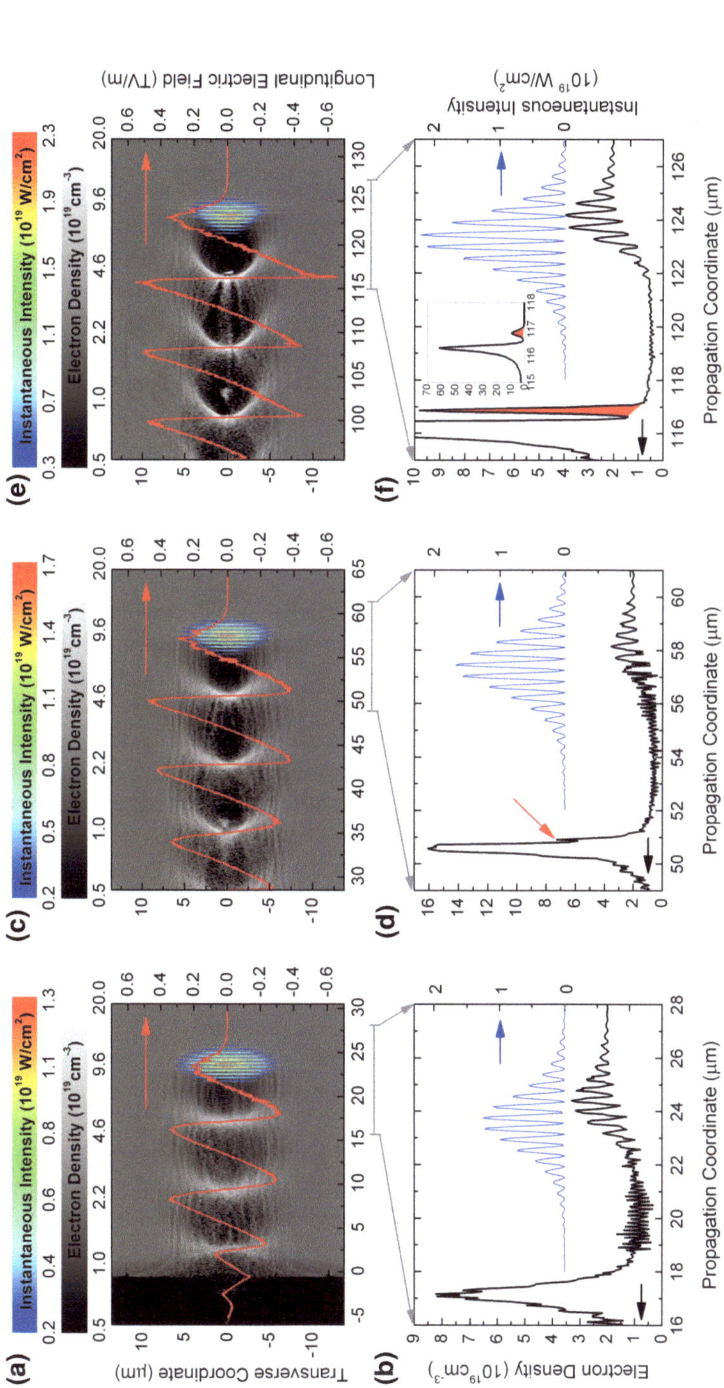

Fig. 5.2 PIC Simulation Results. Shown is the physical state of the accelerator after 23 (**a, b**), 57 (**c, d**) and 123 (**e, f**) μm propagation. Plotted are plasma density (grey-scale contour, *black line*), instantaneous laser intensity (rainbow contours, *blue line*), longitudinal electric field (*red line*). The *red arrow* marks the first injected electrons, the *red peak* is the injected and accelerated electron bunch. The simulations were performed by Michael Geissler with the code ILLUMINATION [33]

wave close to breaking without any prior self modulation. After 25 μm propagation, the laser pulse is still retaining its original spot size and shape. The rainbow-color contour (a) shows the instantaneous intensity (no cycle-average) of the laser pulse, clipped at the $1/e^2$-contour line, the blue line in (b) shows the on-axis line-out of it.

The middle panel (c), (d) shows the system after approximately 57 μm propagation. The laser pulse now exhibits a modified shape: through relativistic self-focusing it has now a 10% smaller spot size and also the Gaussian shape is altered to a somewhat conical one. The plasma electrons need roughly one laser period to acquire relativistic energies, therefore the leading edge experiences no self-focusing but towards the end of the laser pulse this effect has increasing influence on the laser, thereby contracting the pulse-tail. Since the wave did not break so far, also the energy transfer from the laser pulse to the plasma was up to that point only minor. Therefore the self-focusing leads to an intensity-increase from $1.3 \times 10^{19}\,\mathrm{W/cm^2}$ at the beginning to $1.7 \times 10^{19}\,\mathrm{W/cm^2}$ after 57 μm propagation, corresponding to a rise by 33% (d). At this stage, the plasma wave is so strongly driven that the density peak at the back of the first wave trough is almost an order of magnitude larger than the background density. This triggers the onset of the breaking of the wave, the red arrow in figure (d) marks a small density peak that corresponds to the first injected electrons. The peak longitudinal electrical field is approximately 0.4 TV/m, two orders of magnitude smaller than the cold one-dimensional wave breaking limit of 47 TV/m (5.85). Therefore it can be concluded that the wave breaking observed here is different from the strictly longitudinal wave breaking that occurs in 1D. Indeed, inspection of particle tracks shows, that the wave breaking here is transversal: the electrons that have been expelled by the laser pulse are pulled back to the axis by the strong charge separation field built up by the ions in the bubble. They swing back towards the axis and cross it at the back of the bubble, forming a vertex there. At the vertex, a strong electron density spike arises, producing a strong repelling field for the incoming electrons. A fraction of these electrons are then scattered by this field into forward direction, thereby acquiring the necessary speed to get trapped by the wave. This injection process continues until beam loading prevents any further injection: the electric field of the injected electron bunch becomes so strong that it prevents any more electrons from being scattered into the bubble. However, this point is not reached in the present simulation.

The last panel (e), (f) shows the accelerator 25 μm before the pulse leaves the plasma. The plasma wave is now strongly anharmonic, the electrons are now almost completely expelled from the positively charged "bubble" that trails the laser pulse. As the electrons swing back towards the axis, they form a compressed electron sheath that "wraps" the bubble thereby producing a charge distribution that generates the almost perfect sawtooth profile of the longitudinal electric field. Marked in red in the lower panel (f) is the injected electron bunch that has already acquired an energy of 23 MeV. No dephasing (the dephasing length according to Eq. 5.87 is 650 μm) can be observed so far, the electron bunch still is located at the

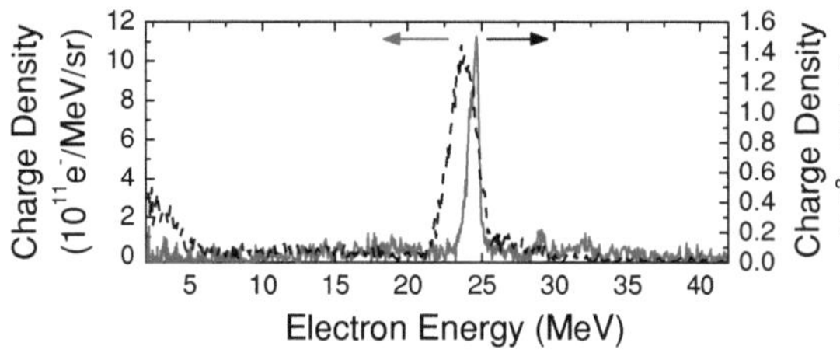

Fig. 5.3 Energy spectrum in the PIC simulation after the electrons propagated out of the plasma several 100 μm (*dashed, black*). Typical experimental electron spectrum (*solid, grey*)

back of the bubble experiencing an accelerating field strength of approximately 0.45 TV/m. The laser pulse at this stage is self-focused down to 75% of its original spot size, its intensity has risen to $2.2 \times 10^{19}\,\mathrm{W/cm^2}$, the conical pulse shape produced by the intensity-dependent relativistic self-focusing effect is now clearly visibly. Laser pulse and accelerated electron bunch are perfectly spatially separated at all times thereby ensuring that all observed properties of the accelerated electron bunch are directly related to the properties of the bubble regime without any modification due to direct laser-electron interaction [34]. The density depletion in the bubble now amounts to 25% corresponding to a electron density of $0.5 \times 10^{19}\,\mathrm{cm^{-3}}$. The injected bunch has an electron density of almost five times the background density and is strongly confined in space and time having a duration of less than one fs at this point. The inset shows the electron density at the back of the bubble to the full scale of the peak in the vertex. Its density is $6.2 \times 10^{20}\,\mathrm{cm^{-3}}$, more than 30 times the background density. The red peak again is the injected electron bunch, its amplitude dwindled by the peak at the bubble vertex. This indicates that the wave structure in principle would support much more injected charge before beam loading becomes relevant. This is also supported by the fact that the accelerating longitudinal electric field is hardly perturbed by the injected bunch.

The electron spectrum after the electron bunch left the plasma and propagated several 100 μm into the surrounding vacuum is given in Fig. 5.3 in black. It shows a monoenergetic peak at 24 MeV which implies together with the accelerating field of 0.45 TV/m an effective acceleration length of about 60 μm. The electron bunch emerging from the plasma carries a charge of about 4.5 pC has a duration on the order of 1 fs and is accompanied by a small exponential background. The ultrashort electron bunch duration is approximately preserved within the range of simulated propagation extending several 100 micrometers behind the gas jet. For comparison, also a typical experimental electron spectrum is shown in grey. Interestingly its spectral width is even smaller than the one produced by simulation.

The simulations show that by using sub-10-fs laser pulses it is possible to access the bubble regime with a pulse energy of only 35 mJ. As shown above, this laser pulse is not able to drive the wave into breaking right from the start of the interaction but is able to do so after self-focusing to 90% of its original spot size which takes place over a propagation distance of approximately 57 μm. Then, the laser intensity is just high enough to produce self-injection but still low enough to allow the wake field to extend over several oscillations before it decays entirely. Indeed, the simulations show that our current laser parameters are close to the threshold below which no self-injection and formation of a stable accelerating structure occurs any more [33]. This is in good agreement with both the analytic theory of bubble acceleration [30] which—for our pulse duration—predicts the onset of the process at pulse energies of about 30 mJ (see Eq. 5.98), and the experimental finding that gradual reduction of the laser pulse energy results in rapid increase of fluctuations of the electron beam properties with acceleration ceasing completely for pulse energies below 25 mJ. This implies that with our current on-target energy of 40 mJ we operate the bubble accelerator near the boundary of its operational regime. Higher driving pulse energies should not only improve stability but—according to our simulations—also dramatically increase the laser-to-electron energy conversion efficiency, from currently $\sim 1\%$ to up to 20%. Furthermore the simulations show that after the pulse diameter has reached its minimum due to self-focusing, the intensity of the laser pulse rapidly shrinks due to pulse depletion and diffraction. Because of this, the laser pulse looses its ability to sustain a stable bubble already after 250 μm propagation. This constitutes another reason to assume that more energetic driver pulses are likely to improve the shot to shot stability of the accelerator.

References

1. Batani, D. et al. (eds.): Atoms, Solids, and Plasmas in Super-Intense Laser Fields. Kluwer, Dordrecht (2001)
2. Jackson, J.D.: Classical Electrodynamics, 3rd edn. Walter de Gruyter, Berlin (2002)
3. Moore, C.I., Knauer, J.P., Meyerhofer, D.D.: Observation of the transition from thomson to compton scattering in multiphoton interactions with low-energy electrons. Phys. Rev. Lett. **74**(13), 2439–2442 (1995)
4. Woodward, P.M.: A method of calculationg the field over a plane. J. Inst. Electr. Eng. **93**, 1554–1558 (1947)
5. Lawson, J.D.: Lasers and accelerators. IEEE Trans. Nucl. Sci. **26**(3), 4217–4219 (1979)
6. Esarey, E., Sprangle, P., Krall, J.: Laser acceleration of electrons in vacuum. Phys. Rev. E. **52**(5), 5443–5453 (1995)
7. Salamin, Y.I., Mocken, G.R., Keitel, C.H.: Electron scattering and acceleration by a tightly focused laser beam. Phys. Rev. ST Accel. Beams **5**(10), 101–301 Oct(2002)
8. Bauer, D., Mulser, P., Steeb, W.H.: Relativistic ponderomotive force, uphill acceleration, and transition to chaos. Phys. Rev. Lett. **75**(25), 4622–4625 (1995)
9. Startsev, E.A., McKinstrie, C.J.: Multiple scale derivation of the relativistic ponderomotive force. Phys. Rev. E **55**(6), 7527 (1997)
10. Decoster, A.: Nonlinear travelling waves in a homogeneous cold collisionless plasma. Phys. Rep. **45**(5), 285 (1978)
11. Gibbon, P.: Short Pulse Laser Interactions with Matter. Imperial College Press, London (2005)
12. Siegman, A.E.: Lasers. 3rd edn. University Science Books (1986)
13. Litvak, A.G.: Finite-amplitude wave beams in a magnetoactive plasma. Sov. Phys. JETP **30**, 344 (1970)

14. Max, C.E., Arons, J., Langdon, A.B.: Self-modulation and self-focusing of electromagnetic waves in plasmas. Phys. Rev. Lett. **33**(4), 209–212 (1974)
15. Sprangle, P., Esarey, E., Ting, A., Joyce, G.: Laser wakefield acceleration and relativistic optical guiding. Appl. Phys. Lett. **53**(22), 2146–2148 (1988)
16. Sun, G.-Z., Ott, E., Lee, Y.C., Guzdar, P.: Self-focusing of short intense pulses in plasmas. Phys. Fluids **30**(2), 526–532 (1987)
17. Borisov, A.B., Borovskiy, A.V., Korobkin, V.V., Prokhorov, A.M., Rhodes, C.K., Shiryaev, O.B.: Stabilization of relativistic self-focusing of intense subpicosecond ultraviolet pulses in plasmas. Phys. Rev. Lett. **65**(14), 1753–1756 (1990)
18. Gibbon, P., Monot, P., Auguste, T., Mainfray, G.: Measurable signatures of relativistic self-focusing in underdense plasmas. Phys. Plasmas **2**(4), 1305–1310 (1995)
19. Pukhov, A., Meyer-ter Vehn, J.: Relativistic magnetic self-channeling of light in near-critical plasma: Three-dimensional particle-in-cell simulation. Phys. Rev. Lett. **76**(21), 3975–3978 (1996)
20. Tzeng, K.-C., Mori, W.B., Decker, C.D.: Anomalous absorption and scattering of short-pulse high-intensity lasers in underdense plasmas. Phys. Rev. Lett. **76**(18), 3332–3335 (1996)
21. Tzeng, K.-C., Mori, W.B.: Suppression of electron ponderomotive blowout and relativistic self-focusing by the occurrence of raman scattering and plasma heating. Phys. Rev. Lett. **81**(1), 104–107 (1998)
22. Oliveirae Silva, L., Mendonça, J.T.: Kinetic theory of photon acceleration: Time-dependent spectral evolution of ultrashort laser pulses. Phys. Rev. E **57**(3), 3423–3431 (1998)
23. Sheng, Z.M., Meyer ter Vehn, J.: Relativistic wave breaking in warm plasmas. Phys. Plasmas **4**(2), 493–495 (1997)
24. Faure, J., Glinec, Y., Pukhov, A., Kiselev, S., Gordienko, S., Lefebvre, E., Rousseau, J.-P., Burgy, F., Malka, V.: A laser–plasma accelerator producing monoenergetic electron beams. Nature **431**, 541 (2004)
25. Geddes, C.G.R. et al.: High quality electron beams from a laser wakefield accelerator using plasma-channel guiding. Nature **431**, 538 (2004)
26. Mangles, S.P.D., Murphy, C.D., Najmudin, Z., Thomas, A.G.R., Collier, J.L., Dangor, A.E., Divall, E.J., Foster, P.S., Gallacher, J.G., Hooker, C.J., Jaroszynski, D.A., Langley, A.J., Mori, W.B., Norreys, P.A., Tsung, F.S., Viskup, R., Walton, B.R., Krushelnick, K.: Monoenergetic beams of relativistic electrons from intense laser plasma interactions. Nature **431**, 535 (2004)
27. Pukhov, A., Meyer-Ter-Vehn, J.: Laser wake field acceleration: the highly non-linear broken-wave regime. Appl. Phys. B **74**, 355 (2002)
28. Lu, W., Huang, C., Zhou, M., Mori, W.B., Katsouleas, T.: Nonlinear theory for relativistic plasma wakefields in the blowout regime. Phys. Rev. Lett. **96**(16), 165002 (2006)
29. Tsung, F.S., Lu, W., Tzoufras, M., Mori, W.B., Joshi, C., Vieira, J.M., Silva, L.O., Fonseca, R.A.: Simulation of monoenergetic electron generation via laser wakefield accelerators for 5–25 tw lasers. Phys. Plasmas **13**(5), 056708 (2006)
30. Gordienko, S., Pukhov, A.: Scalings for ultrarelativistic laser plasmas and quasimonoenergetic electrons. Phys. Plasmas **12**, 043109 (2005)
31. Pukhov, A., Gordienko, S.: Bubble regime of wake field acceleration: similarity theory and optimal scalings. Phil. Trans. R. Soc. A **364**, 623 (2006)
32. Lu, W., Tzoufras, M., Joshi, C., Tsung, F.S., Mori, W.B., Vieira, J., Fonseca, R.A., Silva, L.O.: Generating multi-gev electron bunches using single stage laser wakefield acceleration in a 3d nonlinear regime. Phys. Rev. ST Accel. Beams **10**(6), 061301 (2007)
33. Geissler, M., Schreiber, J., Meyer-Ter-Vehn, J.: Bubble acceleration of electrons with few-cycle laser pulses. New J. Phys. **8**, 186 (2006)
34. Gahn, C., Tsakiris, G.D., Pukhov, A., Meyer-terVehn, J, Pretzler, G., Thirolf, P., Habs, D., Witte, K.J.: Multi-mev. electron beam generation by direct laser acceleration in high-density plasma channels. Phys. Rev. Lett. **83**(23), 4772–4775 (1999)

Chapter 6
Experimental SetUp

6.1 The Light Source: Light Wave Synthesizer 10

It was mentioned in the introduction that many of the milestones in laser-driven electron acceleration were ultimately reached by the ongoing development of ever more powerful driving lasers. In 2004, when three independent groups showed for the first time the generation of mono-energetic electron beams by laser-driven plasma waves [1–3], this was essentially made possible by the availability of high power ultra short pulse lasers allowing for the first time to resonantly excite a plasma wave at densities in the range of $10^{18}\,\mathrm{W/cm^2}$. Here, resonantly signifies that the laser pulse fits longitudinally within one plasma wavelength. This increases the efficiency of wake field excitation and effectively eliminates parametric instabilities like stimulated Raman and Brillouin scattering that are dominant with longer pulses. Figure 1.2 in the introduction, shows the most important experimental results concerning laser driven electron acceleration starting in 2004. The black line shows the limit of the bubble regime as obtained from (4.102). It is today a more or less accepted fact that ultimately always a bubble-like structure is responsible for injection and acceleration of electrons but still the majority of the experiments are clearly out of this regime. In these cases self-compression and self-focusing transforms the laser pulse into the required domain. Only two experiments have so far accessed the bubble regime directly by initial laser parameters, one with a Ti:sapphire based chirped pulse amplifier (CPA) producing pulse powers of more than 100 TW [4] and the other one being the present work, using 5 TW few-cycle pulses.

The ultrashort light pulses used in the present experiments were produced by a novel light source, Light Wave Synthesizer 10 (LWS-10) [5]. Only through the unique properties of this two-staged non-collinear optical parametric chirped pulse amplifier (NOPCPA) was it possible to access the bubble regime with on-target pulse energies of 40 mJ only. A sketch of the layout of LWS-10 is shown in Fig. 6.1. A broadband Ti:sapphire oscillator (Rainbow, Femtolasers GmbH) simultaneously seeds the Nd:YAG pump laser (EKSPLA) and the signal

K. Schmid, *Laser Wakefield Electron Acceleration*, Springer Theses,
DOI: 10.1007/978-3-642-19950-9_6, © Springer-Verlag Berlin Heidelberg 2011

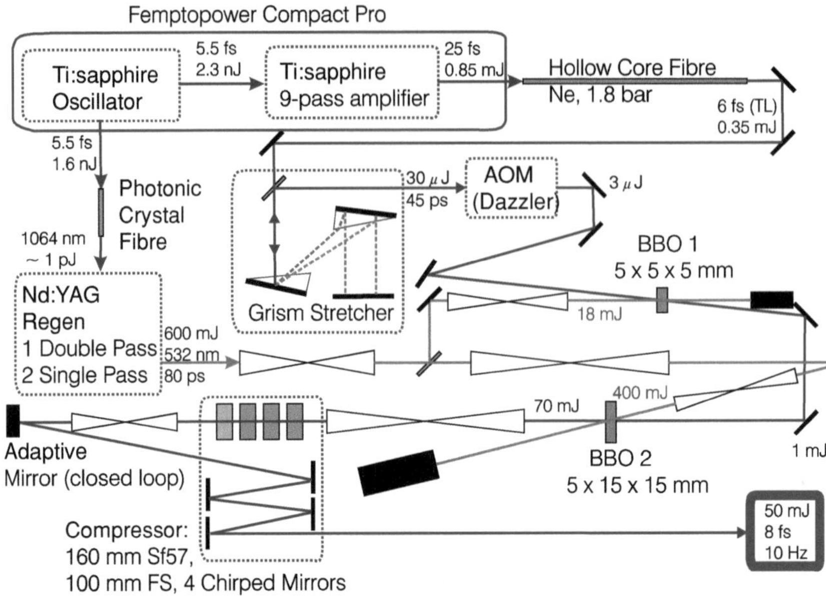

Fig. 6.1 Layout of Light Wave Synthesizer 10 (LWS-10)

amplification chain thereby ensuring optimal synchronization between pump and signal. The oscillator produces pulses with a spectral bandwidth of 300 nm, ranging from 650 to 950 nm, a pulse duration of 5.5 fs and 4 nJ energy at a repetition rate of 80 MHz, 40% of the output seeds the pump laser, 60% the signal amplifier. In order to generate the seed for the pump laser at 1,064 nm from the Ti:sapphire oscillator, the ultrashort pulses from the oscillator are sent through a photonic-crystal fibre where the central wavelength is upshifted to the desired range via soliton-based Raman shift. In this way, 4 pJ out of the 1.6 nJ (0.25%) are converted to the 1,064 nm range. The pump laser is a commercial, flashlamp pumped Nd:YAG amplifier (EKSPLA) that consists of a regenerative amplifier followed by a double-pass and two single-pass stages. After type-II second harmonic generation in a DKDP crystal, it produces pulses with 600 mJ energy and 80 ps (FWHM) duration at a repetition rate of 10 Hz.

The signal amplification chain starts with a commercial Ti:sapphire-based 9-pass amplifier (Femtopower Compact PRO CE, Femtolasers GmbH) that picks oscillator pulses at 1 kHz repetition rate, stretches and amplifies them in a multi-pass amplifier to an energy of 1.2 mJ. Gain narrowing strongly reduces the signal bandwidth such that after compression by a prism-based compressor a pulse duration of 25 fs with 0.85 mJ energy is reached. These pulses are subsequently focused onto the entrance of a 1-m long hollow core fibre (HCF) with a diameter of 250 μm that contains neon gas at a pressure of 1.8 bar. The hollow core fibre prevents the laser from diffracting and thereby ensures high laser intensity over the entire length of the fibre, greatly exceeding the Rayleigh range. At these intensities

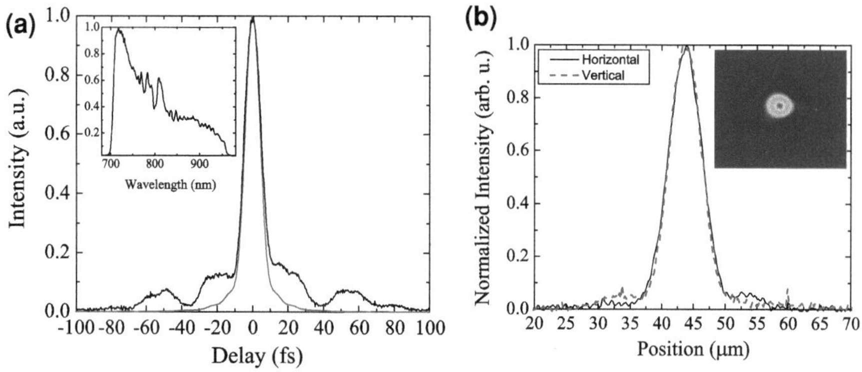

Fig. 6.2 **a** Measured pulse duration of LWS-10 by second-order single-shot autocorrelation (*black*) and fourier transform of the pulse spectrum (*red*). The de-convoluted pulse duration is 8.02 ± 0.12 fs, which is within 5.5% of the fourier limit. The *inset* shows the spectrum of the amplified pulses. **b** Typical focal spot obtained with an F/6 focusing off-axis parabolic mirror. The beam diameter on the parabola is 50 mm, the effective focal length of the parabola is 312 mm. A gaussian fit to the curves (not shown for clarity) yields a horizontal spot diameter (FWHM) of 6.3 μm and a vertical one of 5.9 μm. $85 \pm 5\%$ of the energy are within the first airy-ring. Pulse duration measurement courtesy of Raphael Tautz *a* pulse duration of LWS-10 *b* Focal spot of LWS-10

of $10^{13}-10^{14}$ W/cm^2, neon exhibits a significant Kerr-nonlinearity that leads to strong self-phase modulation of the laser pulses producing a broad output spectrum ranging from 500 to 1,000 nm at a pulse energy of 0.35 mJ. Subsequently, a negative-dispersion reflection grism pair and an acousto-optic programmable dispersive filter (Dazzler, Fastlite) stretch the pulse to 25 ps duration. Grism-stretcher and a Dazzler have each an efficiency of 10% leading to a final seed pulse energy for the first NOPCPA stage of 3 μJ. The Dazzler is also used to pick the 10 Hz seed pulse train from the 1 kHz pulse train coming from the multi-pass amplifier. The first NOPCPA stage consists of a $5 \times 5 \times 5$ mm type-I BBO crystal that is pumped by 15-mJ pulses at 532 nm and amplifies the seed pulses to approximately 1 mJ. The second stage is operated in saturation and consists of a $15 \times 15 \times 5$ mm type-I BBO crystal pumped by 400-mJ pulses. The amplified stretched signal pulses after the second stage have an energy of 70 mJ corresponding to an pump-to-signal conversion efficiency of 18%. After increasing the signal beam diameter, bulk glass (160 mm SF57, 100 mm fused silica) and chirped mirrors are used for compression of the pulses. Losses in the bulks and on the silver mirrors in the compressor setup lead an efficiency of the compression of 70%, yielding compressed pulses with 50-mJ energy, 8 fs duration and spectra covering the range of 700–980 nm (see Fig. 6.2a at a 10-Hz repetition rate. Near-transform-limited performance is achieved by high-order dispersion control using the Dazzler. A pulse duration measurement by a home-made all-reflective second-order single-shot intensity auto-correlator [6] is shown in Fig. 6.2a giving a de-convoluted pulse duration of 8.02 ± 0.12 fs. This is within 5.5% of the fourier limit.

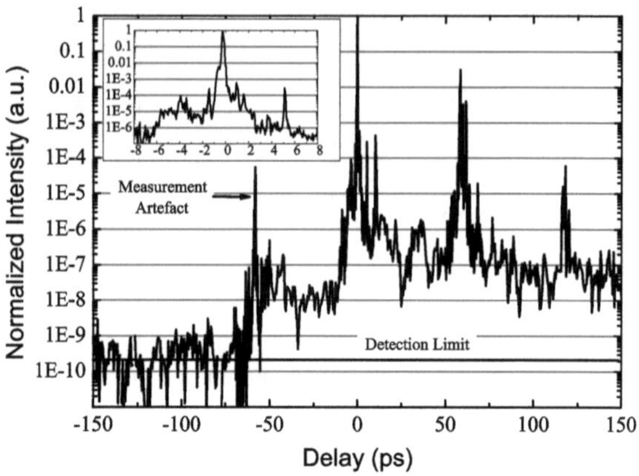

Fig. 6.3 Measured contrast of LWS-10. The measurement was done by third-order autocorrelation using a high-dynamic range autocorrelator constructed by the author. The noise floor of the device in this case is at a contrast ratio of 10^{-10}, with respect to the main peak. The *inset* shows a magnification of the correlation trace close to the main peak

An adaptive mirror (Imagine Optics) in closed loop mode is used to correct the wavefront aberrations and also allows to modify the focal spot diameter without changing the focusing optics in the experiment. A typical focal spot obtained by F/6 focusing during the experiments is shown in Fig. 6.2b. It has an average (with respect to the vertical and horizontal axis) diameter of 6.1 μm corresponding to a peak intensity of up to $1 \times 10^{19}\,\mathrm{W/cm^2}$.

The temporal contrast of the pulses was characterized with a third-order correlator developed by the author [7, 8]. A measurement is presented in Fig. 6.3. The measured contrast ratio with respect to the main pulse is 10^{-5} at a delay of -5 ps, 10^{-8} at -15 ps and better than 10^{-10} at delays smaller than -60 ps. Here, a contrast ratio of 10^{-10} corresponds to the measurement limit of the measurement device. The post-pulses at a delay of 60 ps stem from double internal reflection in the BBO-crystals of the NOPCPA stages. Since they come after the main pulse, they do not influence the experimental conditions but their energy content strongly depends on the delay between pump and seed pulses. A deviation of 10 ps from the optimum alignment is enough to make them almost equally strong than the main pulse thereby constituting a major source of energy loss—the total energy as measured by a power meter is the same in all the cases. Therefore, the pulse-contrast measurement had to be done on a daily basis in order to ensure that the energy of the post pulses was negligible with respect to the main pulse.

The present electron acceleration experiments are the first experimental application of the entirely new light source LWS-10. As such, it did not come as a surprise that the system suffered from fair amount of teething troubles by the time that the major part of the system development including pump-seed synchronization, pump-beam transport, and seed pulse dispersion management had been finished. The three most significant problems turned out to be the tedious optimization procedure for pulse-compression, the low and unreliable temporal contrast of the amplified pulses, and the unpredictable wave front curvature of the main beam that strongly impacted the focusability. Therefore, the first major system upgrade and improvement was indispensable before first light could be sent to the experimental chamber. Significant contributions in the course of this upgrade, mainly concerning the pulse contrast and the focusability, were made in the course of work of the present doctoral thesis. Before experiments could start, within roughly one year's time, the laser system was equipped with the new GRISM-based strecher (this was done by Franz Tavella [5]), the new front-end was installed and one NOPCPA-stage was removed (initially the seed amplification chain consisted of one broadband oscillator and three NOPCPA amplification stages) and an adaptive mirror was installed for wavefront optimization. With all these improvements made, it was finally possible to achieve the experimental results which are presented in the following.

6.2 Setup of the Experiment

The electron acceleration experiments were performed in the radiation protection bunker at Max-Planck-Institute of Quantum Optics (MPQ). The bunker resides on the ground floor whereas LWS-10 is located in a clean-room on the first floor. A remotely controllable vacuum beam line system connects laser and experiment allowing to send the compressed laser pulses to several experimental chambers. Due to the large bandwidth of the pulses produced by LWS-10, silver mirrors have to be employed in the beam line instead of the dielectric mirrors that are commonly used for Ti:sapphire lasers. Seven of these silver mirrors in the beam line plus three more in the experimental chamber unfortunately lead to a rather poor transmission efficiency of approximately 84% not including the off-axis parabola in the experimental chamber. This leads to an on-target energy of 40 mJ that was obtained in the present experiments.

The experiments were performed in a dedicated experimental chamber that includes not only the setup for electron acceleration but also a custom designed electron spectrometer. A photograph of the chamber is shown in Fig. 6.4, a schematic drawing in Fig. 6.5. The laser pulses are focused by an off-axis parabolic mirror (OAP) onto the target medium. Two different OAPs are used, providing F/3 (enhanced silver coated, effective focal length 150 mm) and F/6 (gold coated, effective focal length 312 mm) focusing. A typical focal spot obtained with the F/6 OAP is shown in Fig. 6.2b. It has a diameter (FWHM) of 6.1 μm, the calculated peak intensity is $1.0-1.2 \times 10^{19}\,\mathrm{W/cm^2}$ depending on pulse energy. The focal spot obtained with the F/3 OAP was of similar quality having a diameter of 3.5 μm with a calculated intensity of $3.4-3.6 \times 10^{19}\,\mathrm{W/cm^2}$. As mentioned above, the laser system is equipped with an adaptive mirror in closed loop configuration. However, due to wave-front abberations introduced by the out-coupling semi-transparent mirror that sends a small fraction of the beam-power to the wavefront sensor as well as due to the large number of optical elements between the point of optimally flat wavefronts and the OAP in the experimental chamber, the focal spot needed manual optimization on a daily basis. This was achieved by

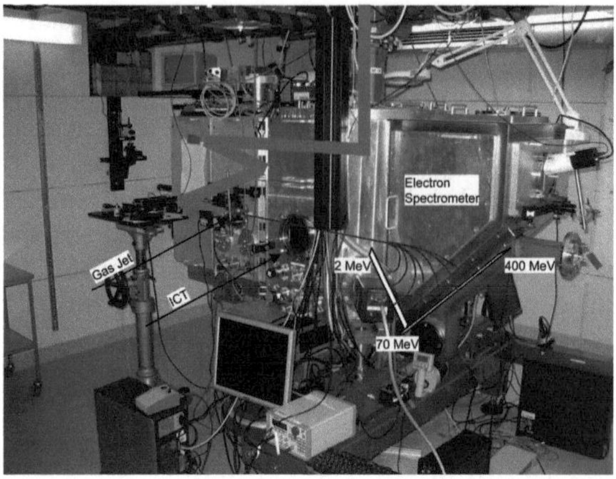

Fig. 6.4 Photograph of experimental chamber

manually setting the adaptive mirror to correct for the additional aberrations causing the quality of the focal spot to change somewhat from day to day. Therefore, the focal spot was characterized before each experimental run using a microscope objective and a beam profiler equipped with a CMOS camera.

The helium interaction medium is provided by a pulsed gas jet. Gas jets of several different kinds—subsonic and supersonic—and sizes were tested, see the section on experimental results for details. The density of all gas jets was measured off line before employment in the experiment, the accessible density range being $10^{18}-10^{20}\,\mathrm{cm}^{-3}$. Details about numeric simulation, design and experimental characterization of the gas jets are given in part I of this work.

The laser-generated plasma channel is imaged transversally by a microscope objective with long object distance onto a CCD camera allowing high-resolution measurements of channel diameter and length. In addition, a weak probe beam can be coupled into the side-view imaging system for assisting in the alignment of the nozzle and for studying the plasma channel.

The electron energy spectrum is measured by a high-resolution focusing permanent magnet spectrometer suitable for analyzing electrons in the range of 2–400 MeV. It comprises a 30−cm × 40−cm focusing permanent magnet with a magnetic field of almost 1 T over a gap of 5 cm. A map of the absolute value of the magnetic field on the center-plane through the gap is shown in Fig. 6.6. As electron detector, 600 highly-sensitive scintillating fibers, three at a given electron energy and coupled to a 16-bit CCD camera are used [9]. Alternatively, a scintillating screen (Kodak Lanex) is imaged to a 12-bit CCD-camera allowing simultaneous measurement of energy spectrum and divergence. While the scintillating fibers are individually shielded against the stray light from laser and plasma, the Lanex-diagnostic only worked when a thin (2 μm) aluminum foil was introduced in front of the spectrometer entrance. The energy calibration was done by particle tracking using the simulated, three-dimensional magnetic field depicted in Fig. 6.6. Measurements of the magnetic field with a hall-probe confirmed the simulation results. The energy resolution of the spectrometer is better than 5%

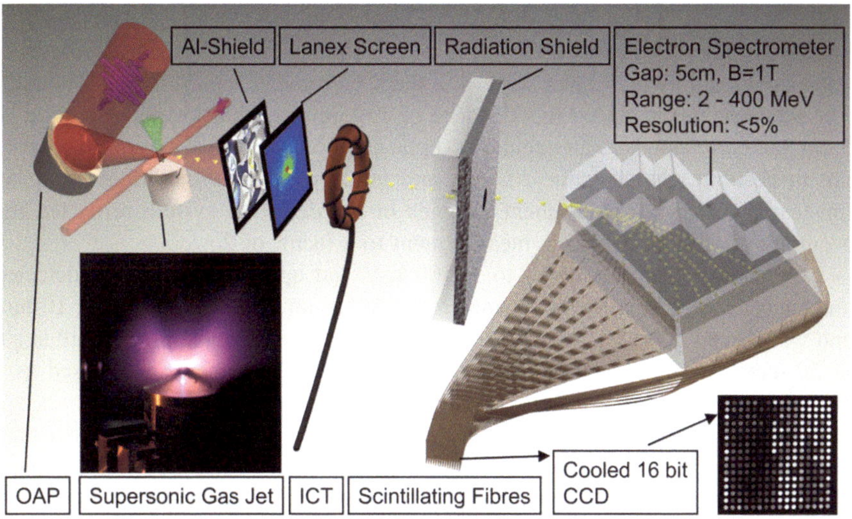

Fig. 6.5 Experimental layout of the laser driven electron accelerator. The laser beam is focused by a gold coated off-axis parabolic mirror onto a helium gas jet. A small fraction of the beam is coupled out by a $1/2''$ mirror and is used as a probe beam. The laser can be blocked after the interaction by a $10\,\mu$m Al-foil allowing to characterize the electron beam (yellow electron bunches) profile on a flippable Lanex screen and the electron beam energy on Lanex screens in the imaging plane of the spectrometer. The Lanex screens were imaged to 12-bit CCD cameras which allowed acquisition at 10 Hz repetition rate; 600 scintillating fibers (in packs of 3) allow to detect the electrons in the imaging plane without blocking the laser beam with the Al-foil and additionally are absolutely calibrated. A highly sensitive, cryogenically cooled 16-bit CCD-camera is used to read out the signal of the fibres. An integrating current transformer (ICT) can be used to measure the charge (>10pC)

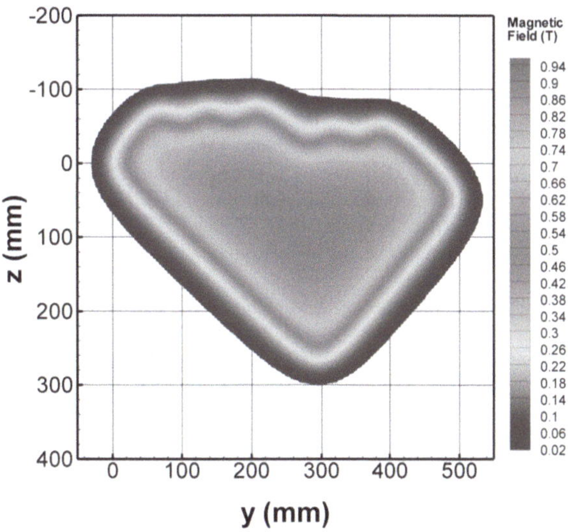

Fig. 6.6 Simulation results of the absolute value of the magnetic field (T) of the electron spectrometer on the center-plane through the gap of the permanent magnet

over the energy range of 3–400 MeV, transversal de-focusing effects lead to increased transversal spot sizes on the detector plane. Therefore, the divergence-values obtained by the Lanex screens in the spectrometer represent upper bounds. The spectrometer is placed in a distance of 790 mm from the gas jet, therefore its 5 cm gap corresponds to an angular acceptance of 63 mrad (full angle). In order to obtain absolute charge values, Lanex and fibres were cross-calibrated using an image plate [18] but subsequent changes in the detection system restricted the accuracy of the bunch-charge measurement to a factor of 2.

Low-energy electrons down to $\sim$100 keV and up to 12 MeV were detected and analyzed by a smaller spectrometer [10]. It has an entrance aperture of 10 mm and was placed 200-mm downstream of the gas jet, covering therefore a full angle of 50 mrad. Lanex screens imaged to two 12-bit CCD cameras were used as a detection system.

As an additional bunch-charge diagnostic an integrating current transformer (ICT, by BERGOZ) was used, but it did not permit the measurement of the typical charges around 10 pC due to the electromagnetic pulse generated by the laser–plasma interaction.

An additional Lanex screen, which could be flipped in and out of the electron beam allowed the characterization of the transverse electron beam profile.

A radiation shield consisting of a 5-cm thick plexiglass plate and 1-cm lead plate were used. The plexiglass contains only atoms with low atomic number (hydrogen and carbon) and therefore stops low energy electrons (approximately up to 5 MeV) and decelerates high energy electrons without producing much Bremsstrahlung. Then the lead stops the remaining high energy electrons.

References

1. Faure, J., Glinec, Y., Pukhov, A., Kiselev, S., Gordienko, S., Lefebvre, E., Rousseau, J.-P., Burgy, F., Malka, V.: A laser–plasma accelerator producing monoenergetic electron beams. Nature **431**, 541 (2004)
2. Geddes, C.G.R. et al.: High quality electron beams from a laser wakefield accelerator using plasma-channel guiding. Nature **431**, 538 (2004)
3. Mangles, S.P.D., Murphy, C.D., Najmudin, Z., Thomas, A.G.R., Collier, J.L., Dangor, A.E., Divall, E.J., Foster, P.S., Gallacher, J.G., Hooker, C.J., Jaroszynski, D.A., Langley, A.J., Mori, W.B., Norreys, P.A., Tsung, F.S., Viskup, R., Walton, B.R., Krushelnick, K.: Monoenergetic beams of relativistic electrons from intense laser plasma interactions. Nature **431**, 535 (2004)
4. Hafz, N.A.M., Jeong, T.M., Choi, I.W., Lee, S.K., Pae, K.H., Kulagin, V.K., Sung, J.H., Yu, T.J., Hong, K.-H., Hosokai, T., Cary, J.R., Ko, D.-K., Lee, J.: Stable generation of gev-class electron beams from self-guided laser–plasma channels. Nat. Phot. **2**, 571 (2008)
5. Tavella, F., Nomura, Y., Veisz, L., Pervak, V., Marcinkevičius, A., Krausz, F.: Dispersion management for a sub-10-fs, 10 tw optical parametric chirped-pulse amplifier. Opt. Lett. **32**(15), 2227–2229 (2007)
6. Tautz, R.: Single-shot characterization of sub-10-fs laser pulses. Diploma Thesis (2008)
7. Schmid, K.: Characterization of ultrashort laser pulses by third order autocorrelation. Diploma Thesis (2004)

8. Tavella, F., Schmid, K., Ishii, N., Marcinkevičius, A., Veisz, L., Krausz, F.: High-dynamic range pulse-contrast measurements of a broadband optical parametric chirped-pulse amplifier. Appl. Phys. B **81**, 753 (2005)
9. Benavides, S.: Spectrometer development for laser-driven electron acceleration. Diploma Thesis (2008)
10. Gahn, C., Tsakiris, G.D., Pukhov, A., Meyer-ter Vehn, J., Pretzler, G., Thirolf, P., Habs, D., Witte, K.J.: Multi-mev electron beam generation by direct laser acceleration in high-density plasma channels. Phys. Rev. Lett. **83**(23), 4772–4775 (1999)

Chapter 7
Experimental Results on Electron Acceleration

7.1 Performance and Stability of the Electron Accelerator

The few-cycle light pulses, having a pulse duration of 8 fs and an energy content of 40 mJ, were focused onto the gas target by F/6 focusing, yielding a focal spot diameter of approximately 6.1 μm FWHM with $85 \pm 5\%$ of the energy being within the airy radius. The exact values of diameter and energy content of the focal spot varies with adaptive mirror settings and performance. During the first experiments, the gas target was provided by a cylindrical (subsonic) nozzle with a diameter of 400 μm allowing to produce densities approaching $10^{21}\,\text{cm}^{-3}$. This eased the initial alignment process and finding the electron signal. At these large densities of more than $10^{20}\,\text{cm}^{-3}$, thermal electron spectra were produced containing typically on the order of 100 pC charge and having a temperature around 4 MeV. This strong signal could easily be detected allowing rough optimization of the nozzle position with respect to the focus and pulse duration as well as verification of the functioning of the diagnostics used in the experiment. The pulse duration was optimized by adjustment of the values for second, third, and fourth order spectral phase on the DAZZLER (see Sect. 6.1). Then, by gradually lowering the pressure and continuously adapting nozzle position and pulse-duration for optimum electron signal, mono-energetic electron spectra could be produced at an electron density of $(2 \pm 0.15) \times 10^{19}\,\text{cm}^{-3}$. Typical monoenergetic electron spectra, as detected by the scintillating fibres (see Sect. 6.2), are shown in Fig. 7.1a, showing the spectra of three different shots. The monoenergetic peaks are located at energies of 13.4, 17.8 and 23 MeV, they contain a charge of approximately 10, 3.5 and 1.6 pC, respectively. In contrast to earlier experiments [1–4] the monoenergetic electron bunches are virtually free of thermal background. The measurement yields meaningful results down to a minimum energy of approximately 2 MeV. Within this limit, a small thermal contribution can be seen, but it is up to a factor of six smaller than the main peak. In order to confirm this result down to even lower energies, another, much smaller, permanent magnet spectrometer was used [5], which allows measurements down to approximately

K. Schmid, *Laser Wakefield Electron Acceleration*, Springer Theses,
DOI: 10.1007/978-3-642-19950-9_7, © Springer-Verlag Berlin Heidelberg 2011

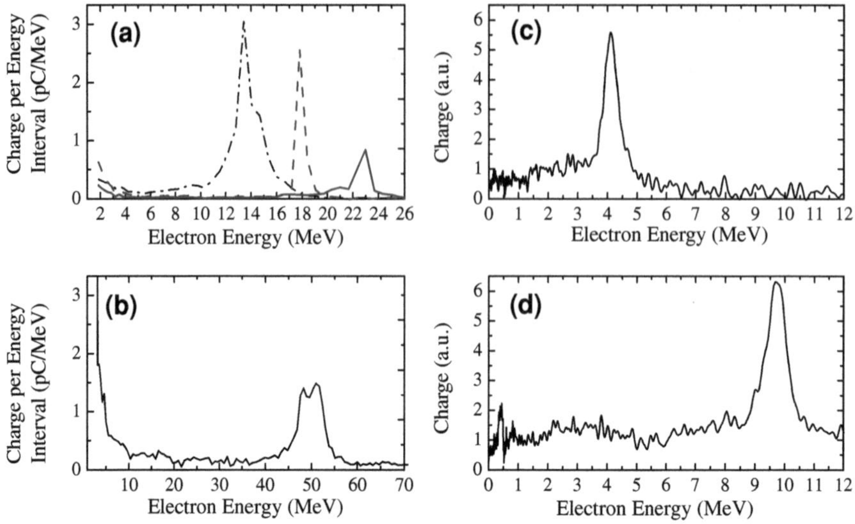

Fig. 7.1 **a** Typical spectra of monoenergetic electron beams from the sub-10-fs laser driven accelerator. They have mean energies of 13.4, 17.8 and 23 MeV. The bunches carry a charge of approximately 10, 3.5 and 1.6 pC, respectively. All three spectra show remarkably few thermal background electrons, an observation that has been confirmed down to energies of 100 keV using a smaller spectrometer (**c**), (**d**). **b** Spectrum with the highest electron energy obtained so far. The spectrum has a double-peaked structure and is centered around 50 MeV. It contains more than 10 pC. All spectra were obtained with a subsonic gas jet produced by a cylindrical nozzle with 400 μm diameter. The laser pulses had an energy of 40 mJ, the spot size was 6.5 μm (FWHM)

100 keV. Two electron spectra obtained with this spectrometer are shown in Fig. 7.1c, d. As detector Lanex was used, which was imaged to a 12-bit CCD camera. The limited sensitivity of the Lanex together with a suboptimal imaging geometry caused by constraints within the vacuum chamber lead to a rather poor signal-to-noise (S/N) ratio. Still, the measurement shows that there is no observable contribution to the monoenergetic electron spectra even at sub-MeV energies. Unfortunately, quantitative comparison with earlier experiments is hampered by the lack of spectral analysis of the thermal electrons in most previous studies [1, 2]. A convenient side-effect of the clean electron beam is the low dose of gamma radiation (Bremsstrahlung) generated when the beam penetrates through the wall of the vacuum chamber.

The three different spectra depicted in Fig. 7.1a were all produced with nominally the same experimental parameters implying that the shot-to-shot reproducibility is rather unsatisfactory. The monoenergetic shot with the highest energy obtained so far is shown in Fig. 7.1b. The mono-energetic feature consists of two narrow, mono-energetic peaks with an energy separation of 3 MeV and a total charge of more than 10 pC. It will be shown below that this multi-bunch structure is a typical feature of the most energetic electron spectra. This spectrum also comprises a considerable thermal background, the maximum amplitude of which is roughly twice as large as the one of the monoenergetic peak. Still, this is orders of magnitude better than in previously realized experiments.

Several possible sources for the strong fluctuations of the electron beam properties are worth considering: first, the above results were obtained in single-shot mode or at low repetition rate, therefore, a slow drift in laser parameters would show up as shot-to-shot fluctuation in the electron beam. To exclude this, the experimental diagnostic tools—most importantly the electron spectrometer—were improved to allow data acquisition at the full 10 Hz repetition rate of LWS-10. The use of optimized microscopic de Laval nozzles produced here a small enough gas load on the vacuum pumps, so that sustained operation at 10 Hz was possible. Since the cooled 16-bit camera used for recording the signal of the scintillating fibres could not be used at 10 Hz, scintillating Lanex screens were introduced into the imaging plane of the electron spectrometer. These then were imaged to fast 12-bit CCD cameras allowing 10 Hz operation as well as simultaneous observation of electron energy and transversal beam profile.

Second, the gas jet emanating from the cylindrical nozzle is strongly divergent and has a gaussian density profile with the peak-density exponentially decreasing with the distance from the nozzle, see Fig. 3.6 for details. Therefore, small changes of the distance between nozzle and laser focus as they occur during alignment lead to large variations of the plasma density making alignment difficult. More importantly, the lack of sharp jet edges will lead to a significant interaction of the laser pulse with the plasma long before the region with high enough density for acceleration is reached thereby altering the laser pulse shape. This in turn could alter the point where wave breaking is reached as well as the subsequent laser–plasma evolution. As the electron bunch then exits the plasma, it can again be modified by the extending gas jet exit gradient. These considerations triggered the use of supersonic gas jets produced by de Laval nozzles. It has been extensively detailed in Part I of this work that only through allowing the gas to assume supersonic flow conditions already *inside* the nozzle, it is possible to produce a gas jet with a flat-top density profile in vacuum. The nozzle producing the gas jet in the experiments had an exit diameter of 300 μm, a technical drawing can be found in Appendix B, Fig. B.2.

A nice shot acquired with this setup is shown in Fig. 7.2. The upper panel is the raw data plotted in false colors, the transversal and energy line out are in calibrated units. The monoenergetic electron bunch has an energy of 24.6 MeV with 3.3% energy spread (FWHM), a divergence of 6.3 mrad (FWHM) and a charge of 3 pC. There is no visible thermal background in the electron spectrum. As mentioned above, the value for the divergence is an upper limit due to the transversally slightly de-focusing effect of the magnetic field of the electron spectrometer. Although the above mentioned experimental modifications did not lead to an entirely stable electron beam, they did improve the shot-to-shot reproducibility enough to conduct meaningful statistical analysis and parameter scans, albeit with rather large statistical error. Figure 7.3 shows a time series of 1,500 electron spectra. Only the shots showing at least one monoenergetic energy peak are shown. In case that more than one peak was present, the one with the largest

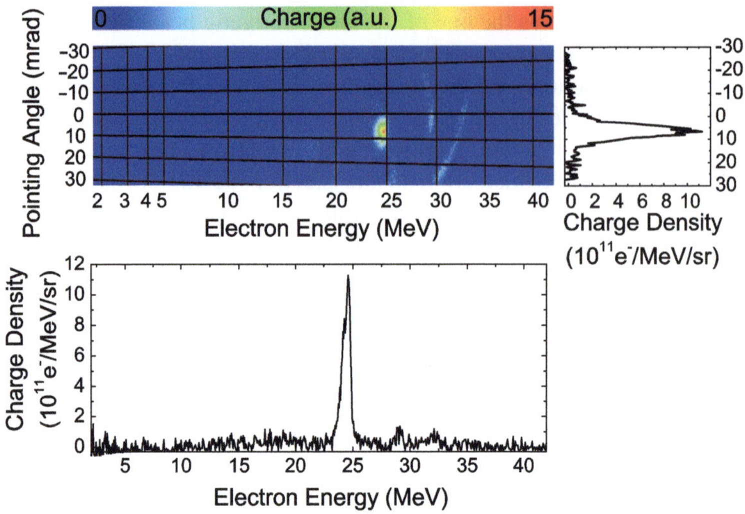

Fig. 7.2 Electron spectrum and transverse beam size obtained with a scintillating screen placed in the electron spectrometer and imaged onto a CCD camera. The false-color plot shows the image on the screen with the right and the bottom panels depicting calibrated line-outs along the energy axis (*lower plot*) and the transversal axis (*right side*). The displayed result was obtained with a 300-μm supersonic gas jet and exhibits a monoenergetic peak at 24.6 MeV with 3.3% energy spread (FWHM), a divergence of 6.3 mrad (FWHM) and a charge of 3 pC. The laser pulses had an energy of 40 mJ, the spot size was 6.5 μm (FWHM)

amplitude was taken. The plot shows the energy of that monoenergetic peak versus shot number. The grey line shows the original data, the black and orange lines represent the FFT—low pass filtered data with a window size of 5 and 25 points, respectively. The stretches of very low yield of monoenergetic spectra between shots 2,400 and 2,600 or between 3,100 and 3,300 are most likely caused by long-term variations of laser parameters which unfortunately were always present during the experiments. A zoom into the shot series between shots 3,550 and 3,800 which shows a constant yield of 50% is shown in the lower panel of the figure. The peak-to-peak variation of the electron energy within this shot range amounts to 100%. Low-pass filtering the data thereby averaging out all variations on a scale shorter than 25 shots yields an almost constant value of 7 MeV (orange line). The upper panel also shows that this low-pass filtered data effectively averages out all statistical fluctuations and shows reliably the slow variation of laser parameters. Applying the same FFT based low pass filter with a window of 5 points gives the black line. Interestingly, it reveals an oscillatory behavior with a peak to peak amplitude of 20% of the average value and with a time constant of 30–40 shots corresponding to 12–16 s, since the data was acquired at 2.5 Hz in this case. The source of this periodic variation is unclear but is also most likely linked to the periodic variation of a laser parameter. The fast, stochastic shot-to-shot fluctuation visible in the grey plot is caused by the acceleration process either through parametric amplification of fluctuations in laser- and gas jet parameters or through processes intrinsic to the injection and acceleration in the plasma wave. To shed light on the dependence of the electron beam properties on the variation of experimental parameters, several parameter scans were conducted. The statistical analysis of a scan of plasma density is shown in Fig. 7.4.

The figure shows the fraction of monoenergetic peaks with respect to the total number of shots (orange) and plots of the charge (upper panel) and energy (lower panel, black squares) of the monoenergetic spectral peak. Each point in the charge and energy plots represent the average value of 300 shots with the error bars representing the root-mean-square error about that average value. The grey line shows the temperature of the thermal electrons in case that maxwellian spectra were produced. The fraction of monoenergetic shots has a sharp maximum around $2 \times 10^{19}\,\mathrm{cm}^{-3}$ and rapidly decays towards higher and lower density. A change of the plasma density of 10% decreases the fraction of monoenergetic shots by a factor of two. The average electron energy at the point of optimum yield is 10 MeV in this case. It was consistently observed in many experiments that slightly increased plasma density leads to higher electron energies but also to decreased yield of monoenergetic spectra and larger shot-to-shot variations. As can be seen in the upper panel, the charge of the monoenergetic electron bunch also varies strongly from shot to shot but does not seem to depend on the plasma density.

PIC simulations and the scaling laws [6] show that the length of the self-guided plasma channel is proportional to the Rayleigh range. For the parameters of the present experiments, the scaling laws predict a channel length of roughly twice the Rayleigh range, thus corresponding to the confocal parameter. Therefore, if pump depletion can be neglected and if the accelerating electric fields are the same, higher electron energies can be expected for longer focal length due the increased acceleration length. Histograms of monoenergetic electron energy for a spot size of 4 and 5.4 μm corresponding to a confocal parameter of 86 and 170 μm respectively, are given in Fig. 7.4. In the case of the longer confocal parameter, a decaying tail towards higher energies is developed.

Obviously, increased plasma density and increased spot size cause similar effects in the measured electron spectra. Due to the large fluctuations in the electron beam, no conclusive distinction in terms of electron beam parameters between the two cases can be given here.

The last presented parameter scan is a systematic variation of the position of the gas jet along the propagation axis of the laser beam. The results are presented in Fig. 7.5, showing absolute and relative yield of monoenergetic electrons and their average energy versus gas jet position along the laser beam propagation axis. Increasing position values correspond to movement of the gas jet towards the focusing off-axis parabolic mirror. Error bars of energy are RMS variation around the average (black line). Since the RMS error is very sensitive to single outliers, the fraction of shots having energies of 20–80% (grey line) and 80–100% of the maximum energy are also shown. The confocal parameter of the laser beam calculated for the measured spot size of 3.5 μm is 70 μm, the diameter of the gas jet

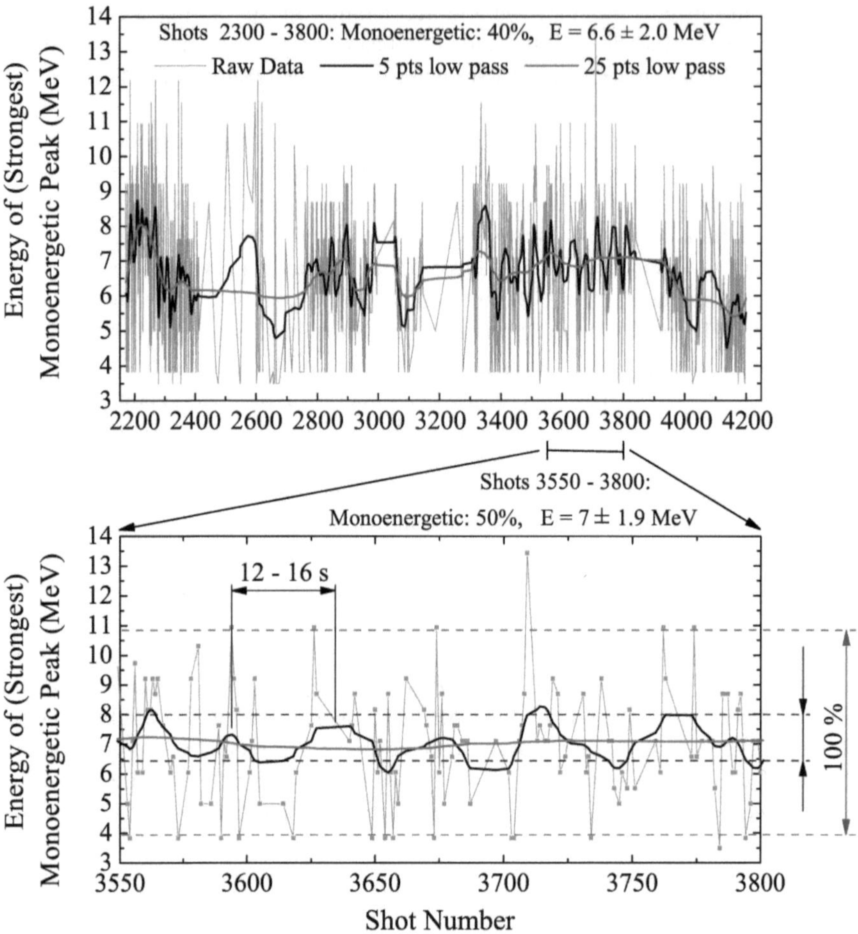

Fig. 7.3 The *upper panel* shows a time series of monoenergetic electron energy. The series contains 1,500 shots out of which 40% are monoenergetic. The majority of the other spectra is non-maxwellian but not monoenergetic, a small fraction is maxwellian. The plots are energy versus shot number, only monoenergetic shots are shown, the others are skipped. The *black* and *orange* lines represent the FFT—low pass filtered data with a window 5 and 25 points, respectively. The *lower panel* shows a magnification of the stretch from shot number 3,550 to 3,800. It shows an increased monoenergetic shot fraction of 50% and seems to be free of long-term variations which most likely are attributed to changes in laser parameters. These results were obtained with the 300 μm de Laval nozzle and 3.5 μm (FWHM) focal spot size

was 300 μm. Electrons are accelerated within a range of 250 μm. The optimum range where the most monoenergetic electron spectra are produced is approximately 50 μm long and lies between −50 and 0 μm. Within this range no significant variation of electron beam parameters with nozzle position is observed. Interestingly, when moving the gas jet away from the focusing parabolic mirror, a peak

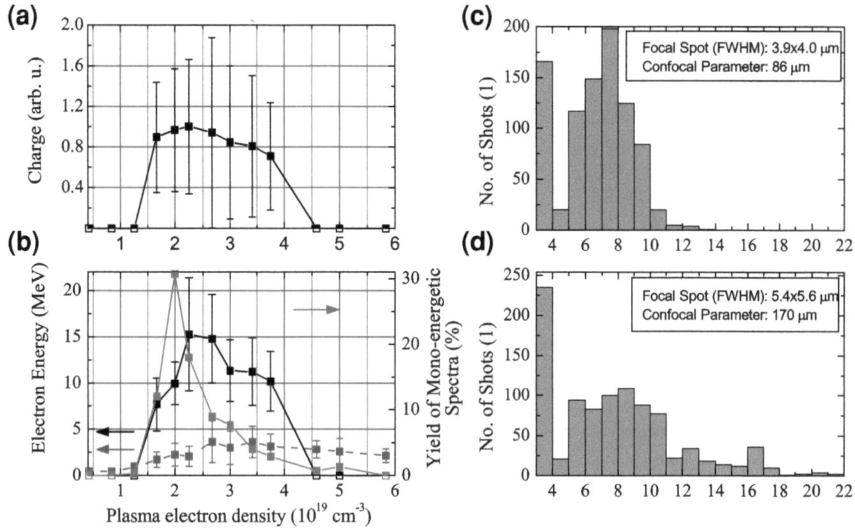

Fig. 7.4 **a**, **b** Results of a density scan with the 300 μm de Laval nozzle and 6 μm (FWHM) focal spot size. The *upper panel* shows the charge in the monoenergetic peak, the *lower panel* the fraction of shots having monoenergetic spectra (*orange*), the average energy of the monoenergetic peak (*black*) and the temperature of the thermal electron spectra (*grey*, *dashed*). Each measurement point corresponds to 300 shots, the *error bars* to the RMS fluctuation about the average. Histograms of monoenergetic electron energy for a spot size of 4 μm **c** and 5.4 μm (**d**). For the larger focal spot corresponding to a longer confocal parameter, a high energy tail shows up. The de Laval nozzle producing the gas jet had a diameter of 300 μm

of almost 80% in the monoenergetic yield is produced at a position of −100 μm just before acceleration ceases completely. The average energy at this position is approximately 7 MeV and, thus, almost half the value that is obtained within the optimum range.

A parameter scan with varying laser energy was also conducted with the main result that electron acceleration only works at on-target energies larger than 25 mJ with optimum stability reached above 30 mJ. Above this value no significant correlation between laser energy and electron beam parameters could be observed. A similar result was obtained for the pulse duration: only within a small range around the optimum pulse duration, electron acceleration was possible at all. A deviation from the optimum group delay dispersion (GDD) of $\pm 50\,\mathrm{fs}^2$ significantly reduced the yield of monoenergetic electron spectra. Assuming that the optimum setting corresponds to the shortest pulse of 8 fs, this deviation in GDD corresponds to a pulse duration of approximately 13 fs.

Under optimal conditions, hence after manual optimization of laser pulse duration, gas jet position, and plasma density, shot series like the one depicted in Fig. 7.6 could be produced.It shows the electron spectra produced by 27 consecutive shots with all of them showing monoenergetic features between 15 and 35 MeV. Due to

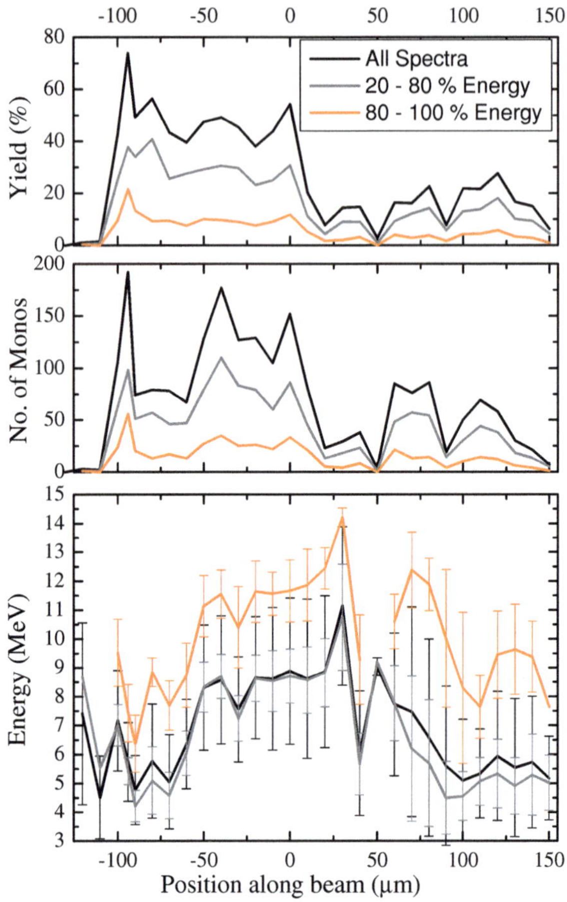

Fig. 7.5 Absolute and relative yield of monoenergetic electron bunches and their average energy versus gas jet position along the laser beam propagation axis. Increasing position values correspond to movement of the gas jet towards the focusing off-axis parabolic mirror. *Error bars* of energy are RMS variation around the average (*black line*). Also the fraction of shots having energies of 20–80% (*grey line*) and 80–100% of the maximum energy are shown. The calculated confocal parameter of the laser beam is 70 µm, the diameter of the gas jet is 300 µm

long-term variation in laser parameters similarly to the behavior shown in the upper panel of Fig. 7.3, the accelerator intermittently dropped in performance leading again to an optimum average yield between 50–80% monoenergetic spectra. Quite obviously, the spectra in Fig. 7.6 differ one from one another in almost all important parameters like charge, energy, and divergence. In order to exclude propagation effects like beam loading, laser pulse depletion and electron bunch dephasing, a much smaller gas jet was used as a target as well. The results of this measurement are shown in Fig. 7.7 showing again two series of consecutive shots. These spectra where measured with the scintillating fibres, so the information about the electron beam divergence is lost. The charge and, hence, the amplitude of the monoenergetic spectral peaks still varies from shot-to-shot but the improvement in terms of reproducibility of the energy spectrum is remarkable. This last result that is shown in this section marks the optimum performance in terms of stability that could be accomplished during the present work.

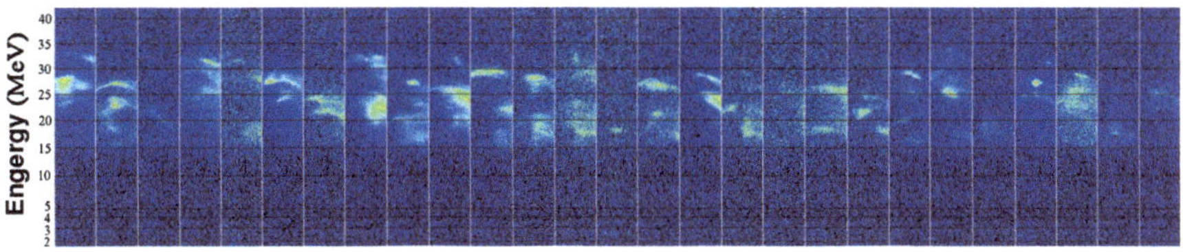

Fig. 7.6 These electron spectra were obtained at 10 Hz repetition rate with a supersonic gas jet with 300 μm diameter. The focal spot size was 6 μm (FWHM). The vertical coordinate in each image is the energy axis, the horizontal coordinate is the transversal (position) axis. Compare to Fig. 7.2 for more information

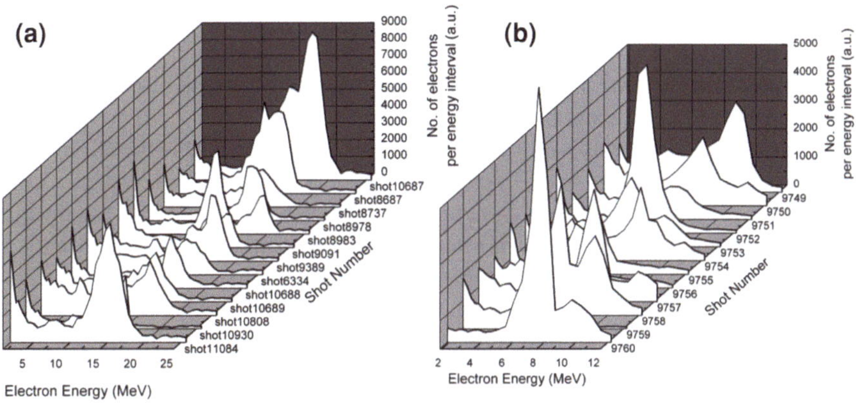

Fig. 7.7 This shot series was obtained at 3.3 Hz repetition rate with a supersonic gas jet with 150 μm diameter. The focal spot size was 6 μm (FWHM), **a** shows all shots (out of 800) with a monoenergetic peak at an energy larger than 10 MeV. Only few shots meet that criterion but these show quite similar spectra, **b** shows one (out of several) consecutive shot series, where each shot shows a monoenergetic peak. Out of 800 shots, 37% are monoenergetic, the average energy of the monoenergetic peak is 7.6 ± 1.6 MeV (RMS)

7.2 Multiple Accelerated Electron Bunches

A feature that was regularly obtained during the experiments are electron spectra containing multiple electron bunches that are well separated in energy and sometimes also in direction. One nice example of such a spectrum is depicted in Fig. 7.8. Six collimated and spectrally narrow electron bunches are easily visible in the energy range between 20 and 30 MeV. The phenomenon of multiple electron bunches is also seen in PIC simulations, it is treated, for example, in [7]. Analysis of the data of tens of thousands of shots showed that predominantly the highest energy spectra show more than one bunch. By filtering out of this large amount of spectra only those which have at least one monoenergetic peak at energies larger than 35 MeV the spectra displayed in Fig. 7.9 are obtained. All of them but one (the second from the left) show multiple bunches and were obtained

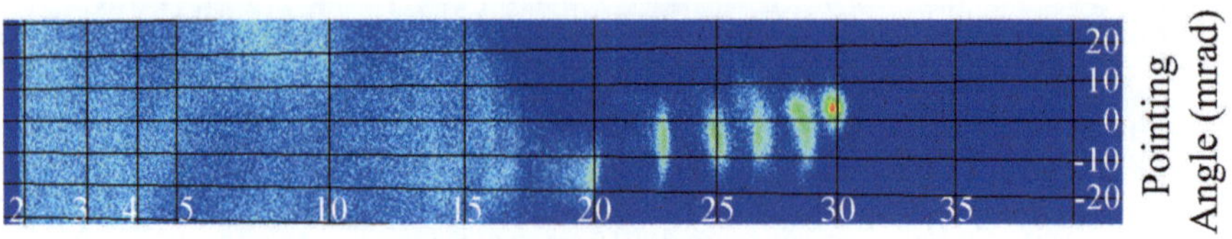

Fig. 7.8 This shot was obtained with a supersonic gas jet with 300 μm diameter and a focal spot size of 12 μm (FWHM). Six clearly separated monoenergetic electron bunches at different energies and with slightly different pointing and divergence are visible. Similar spectra with up to nine bunches were obtained. This type of spectrum was almost exclusively observed with a larger focal spot size (12 μm (FWHM) instead of 6 μm) where up to 3% of the spectra showed more than three peaks

with a spot size of 6 μm. With an enlarged spot size of 12 μm the rate of generation of multi-bunch spectra significantly increased.

In principle two effects—or a combination of them—can lead to the acceleration of more than one electron bunch: consecutive injection of more than one bunch into the same bubble and injection in more than one plasma wave trough with one injected bunch each. In the present case, it is seen in PIC simulations (see Chap. 5.2) that—in contrast to the fully developed bubble regime—the plasma wave does not break down entirely after the first oscillation. Rather a prolonged plasma-wave, extending over several oscillations is trailing the laser pulse and reaches wave breaking in more than one wave trough also in the simulations. Therefore, injection into several plasma wave buckets seem to be the most likely cause of the multi-bunching. Since the fully developed bubble regime only allows for the first bubble to survive the strong wave breaking, the appearance of multiple bunches in the experimental spectra is an indication that the accelerator operates just at the limit of the bubble regime, which also could cause the observed instability. This is confirmed by the fact that the production rate of multiple-bunch spectra increased significantly with increasing laser spot size whereas with smaller focal spot and shorter gas jet, more stable acceleration at lower electron energy and no multi-bunching could be observed.

7.3 Discussion of the Experimental Results

The most important result of this work is the experimental demonstration that electron acceleration is possible with the unprecedentedly low laser pulse energy of only 40 mJ provided that the pulse duration is in the few-cycle-regime. These unique properties of the laser system LWS-10 allowed to consistently satisfy the parametric requirements posed by the scaling laws of the bubble regime. Supported by PIC simulations matching real world experimental conditions, a parameter regime was experimentally delimited within which monoenergetic electron beams can be reliably created. The accelerated electron bunches, depending on laser spot size and plasma density have energies between 5 and 50 MeV, a divergence around

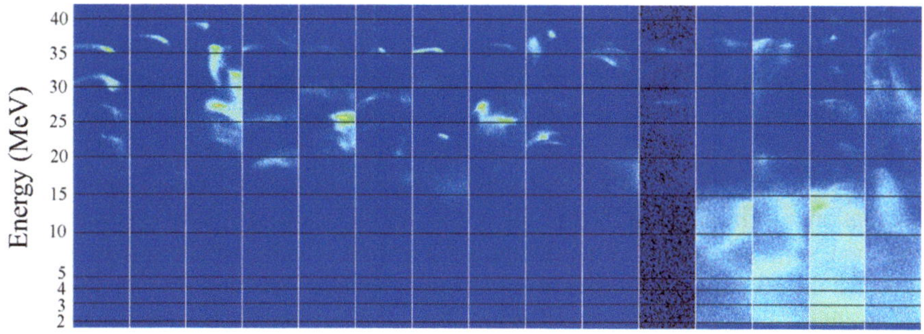

Fig. 7.9 Out of several thousand shots acquired with 300 μm diameter gas jet and 6 μm (FWHM) focal spot size these 15 shots are the only ones with monoenergetic peaks above 35 MeV. As can be seen, they all show multiple peaks

7 mrad, and a charge between 1 and 10 pC. The electron beam is under optimal conditions virtually free of spectral contributions with thermal characteristics. This feature is confirmed down to energies in the 100 keV regime. All these features are well reproduced by PIC simulations.

All relevant parametric dependencies of the acceleration process on experimentally available variables were analyzed. It was found that there is a well-defined optimum density around $2 \times 10^{19}\,\text{cm}^{-3}$ which just fits the requirement of the bubble regime that the longitudinal pulse length should be approximately equal to the radius of the bubble. De-tuning the density by approximately 10% towards higher values leads to larger electron energies but also to larger shot-to-shot fluctuations. This can easily be understood because a larger plasma density will lead to a larger space charge enclosed in the bubble which in turn leads to a larger longitudinal electric field that accelerates the electrons. The larger shot-to-shot fluctuations can be explained either by dephasing, which means that the accelerated electron bunch outruns the driving laser pulse, or by laser pulse depletion which is faster for higher plasma densities. PIC simulations show that in this parameter regime (few-cycle, low energy pulses) laser pulse depletion and intensity loss due to diffraction are the dominant effect. The impact of larger spot size on the electron beam parameters is qualitatively similar to the one of higher plasma density: again, the energy increases together with the shot-to-shot fluctuations. At the largest spot size tested within this work of 12 μm, multiple bunches within one shot occur at a significantly higher rate pointing towards a less strongly driven wake-field that allows for more than one wake field trough (bubble) to develop and subsequently break. Therefore, this constitutes an upper limit of operation of the electron acceleration regarding the focal spot size.

The shot-to-shot reproducibility of the electron beam remains unsatisfactory with optimum results reached using a short plasma length of 150 μm and a spot size of 6 μm. In this case, up to 50% of the produced shots are monoenergetic and show comparable spectra with average energies around 8 to 12 MeV and a RMS energy fluctuation of ±20%. Series of more than ten consecutive shots with monoenergetic electron spectra could be observed in this case. Again, the fact that a short plasma improves the stability points towards laser pulse depletion due to energy transfer and diffraction as a source of instability—for such a short plasma, no visible dephasing occurs in the PIC simulations. It has also been observed that the nice gaussian spot, which is obtained in the focus, rapidly degrades and develops a rich structure as one moves longitudinally out of the focus. This could induce asymmetries in the plasma wave which in turn could cause shot-to-shot fluctuations in the electron beam. Since also in this case no stable operation was obtained, something else seems to alter the electron bunch properties from shot to shot. All the parametric instabilities like stimulated Brillouin and Raman scattering are excluded by the short pulse duration, so that what mainly remains as possible reasons for the fluctuations is first, the strongly anharmonic wake field that depends non-linearly on the laser pulse parameters, and the injection process in the bubble regime. The latter one starts as soon as the transversal wave-breaking threshold is reached and terminates when either the laser pulse is depleted or leaves the plasma or when the space charge of the injected electron bunch prevents any further loading of the bubble. To judge one from the other, parameter scans using full 3D PIC codes must be conducted. Due to limitations in available computing resources, this could not be accomplished within the present work. However, at least beam loading can be excluded as a dominant factor because the simulations show that the electric field caused by the charge of the injected bunch hardly affects the electric field within the bubble at all.

References

1. Yamazaki, A., Kotaki, H., Daito, I., Kando, M., Bulanov, S.V., Esirkepov, T.Zh., Kondo, S., Kanazawa, S., Homma, T., Nakajima, K., Oishi, Y., Nayuki, T., Fujii, T., Nemoto, K.: Quasi-monoenergetic electron beam generation during laser pulse interaction with very low density plasmas. Phys. Plasmas **12**(9), 093101 (2005)
2. Masuda, S., Miura, E., Koyama, K., Kato, S., Adachi, M., Watanabe, T., Torii, K., Tanimoto, M.: Energy scaling of monoenergetic electron beams generated by the laser-driven plasma based accelerator. Phys. Plasmas **14**(2), 023103 (2007)
3. Hsieh, C.-T., Huang, C.-M., Chang, C.-L., Ho, Y.-C., Chen, Y.-S., Lin, J.-Y., Wang, J., Chen, S.-Y.: Tomography of injection and acceleration of monoenergetic electrons in a laser-wakefield accelerator. Phys. Rev. Lett. **96**(9), 095001 (2006)
4. Hidding, B., Amthor, K.-U., Liesfeld, B., Schwoerer, H., Karsch, S., Geissler, M., Veisz, L., Schmid, K., Gallacher, J.G., Jamison, S.P., Jaroszynski, D., Pretzler, G., Sauerbrey, R.: Generation of quasimonoenergetic electron bunches with 80-fs laser pulses. Phys. Rev. Lett. **96**(10), 105004 (2006)
5. Gahn, C., Tsakiris, G.D., Pukhov, A., Meyer-ter Vehn, J., Pretzler, G., Thirolf, P., Habs, D., Witte, K.J.: Multi-mev electron beam generation by direct laser acceleration in high-density plasma channels. Phys. Rev. Lett. **83**(23), 4772–4775 (1999)
6. Gordienko, S., Gordienko, S.: Scalings for ultrarelativistic laser plasmas and quasimonoenergetic electrons. Phys. Plasmas **12**, 043109 (2005)
7. Oguchi, A., Zhidkov, A., Takano, K., Hotta, E., Nemoto, K., Nakajima, K.: Multiple self-injection in the acceleration of monoenergetic electrons by a laser wake field. Phys. Plasmas **15**(4), 043102 (2008)

Chapter 8
Next Steps for Optimizing the Accelerator

As shown in Chap. 7, it is now routinely possible to generate monoenergetic electron bunches with the few-cycle laser-driven electron accelerator. Under optimal conditions and tuned for low energies in the range around 8–12 MeV by using very small gas jets of 150 μm length, the electron-beam is stable enough to be applied to first experiments. However, at higher energies—the accelerator has been shown to be capable of producing up to 50 MeV monoenergetic pulses—the shot-to-shot reproducibility of the electron beam parameters is still unsatisfactory. Figure 7.3 shows that the fluctuations have a systematic and a stochastic component. The systematic component is presumably caused by—at the moment uncontrollable—slow drifts in laser and gas jet parameters. Efforts are under way to reduce this drifts by introducing several active feedback loops that control beam location and direction inside the light amplification chain as well as the timing between pump and seed beam. Furthermore, diagnostics are under development that will allow recording the most important laser pulse parameters as energy, spectrum, and focal spot shape and pointing for every shot. This should not only reveal to which of these parameters the acceleration process is most sensitive but should also allow post selection of the good shots in a data set. At 10 Hz repetition rate—and with the NOPCPA process in principle posing no limitation on the maximum repetition rate, providing a corresponding pump laser exists—it would probably be acceptable to lose 10 or 20% of the shots through post-selection. However, it seems that even with a perfectly stable laser, the much larger stochastic component of the fluctuations would remain.

The route to eliminate the stochastic fluctuations of the electron beam parameters is not obvious because the source of these fluctuations is at the moment unknown. As discussed in Sect. 7.3, comparison of PIC simulations and experiment point towards laser-intensity loss due to pump depletion and diffraction and the injection process as the major contributors to the observed instability. Also the beam-degradation out of the focus cannot be excluded at this point. To avoid or reduce all these effects, the interaction length can be reduced or the laser pulse energy can be increased allowing larger spot size and longer Rayleigh range.

K. Schmid, *Laser Wakefield Electron Acceleration*, Springer Theses,
DOI: 10.1007/978-3-642-19950-9_8,

The first option was successfully tested by using novel, very small supersonic gas jets produced by microscopic de Laval nozzles. The diameter of only 150 μm of this gas jet removed, or at least alleviated, the problem of laser pulse depletion and yielded a more stable electron beam at energies around 10 MeV. However, even in this case it can be seen in Fig. 7.7 that some fluctuation remains. One possible source for these remaining fluctuations could be an intrinsic property of the bubble regime: the self injection mechanism.

In the past, self injection has been extensively used in almost all successful experiments concerning laser driven electron acceleration because it provides a simple way to inject a sizeably number of electrons into the accelerating phase of a plasma wave. The process responsible for self-injection is wave-breaking. It has been extensively described in Sects. 5.1.6 and 5.2. Wave breaking starts as soon as the amplitude of the plasma wave surpasses a threshold value but the point at which this threshold value is reached is not well-defined because its exact location depends on the previous laser pulse evolution in the plasma. Of course, later injection produces less energetic electron bunches if it is assumed that the acceleration length stays constant. Once injection has started, it goes on until the laser pulse leaves the plasma, the laser pulse is depleted or beam loading prevents further injection. This ongoing injection can produce a considerable energy spread of the electron bunch which only can be compensated again if dephasing is reached. Therefore, restricting injection to one well-defined point in the laser–plasma channel seems to be a promising route for improving the stability of the accelerator. This has been experimentally confirmed in [1] and [2]. In these experiments a second, counter-propagating, low energy "injection beam" is overlapped with the main drive-beam at the point where injection should be triggered. The intensity of the main drive-beam is kept below the self-injection threshold so that injection only occurs where the two beams interfere with each other. The results presented in [1] and [2] show a dramatic improvement of shot-to-shot reproducibility using this scheme. Of course, overlapping two laser foci of a high power laser in time and space is not a trivial task and, therefore, an experimentally more straightforward and easy to use method would be more than welcome.

Another concept for triggering injection in a wake field below the wave breaking threshold goes back to a concept presented by Bulanov et al. [3] in 1998. In this work it is theoretically and numerically shown that a downward density transition of sufficient amplitude and finite gradient always leads to sizeable injection of background electrons into the plasma wave excited by an ultra-intense laser pulse. In the case studied in this work, the width of the density gradient covers several plasma wavelength. It was shown in Sect. 5.1.6 that the central prerequisite for injection of background electrons into the accelerating phase of a plasma wave is that the velocity of the plasma electrons becomes larger than the phase velocity of the plasma wave. The presence of a downward (with respect to the direction of propagation of the wake field) density transition now leads to a locally reduced phase-velocity because the wake field gradually adapts its wavelength inside the higher density region to a longer wavelength in the lower density region. In each of the regions, of course, the phase-velocity is given by the group velocity of the laser pulse. Just during the density transition, the phase–velocity is temporarily reduced simply because the plasma wave length gets longer. This decreased phase-velocity now facilitates injection of electrons well below the wave breaking threshold.

Later, this concept was extended by Suk et al. in 2001 [4] to the limit of infinite gradient produced by a step-like density transition for the case of plasma wakefield acceleration (PWFA), which employs an ultra-relativistic electron bunch as driver of the plasma wave. In this case of a step-like transition, electrons that cross the density transition which separates the high density region from the low density region are suddenly re-phased in the wake field due to the phase-jump of the wake field that occurs at the density transition. Further studies of this concept concerning PWFA can be found in [5, 6].

As laser technology improved and LWFA became more popular, injection at density transitions with short transition scale-length was extensively studied in this context as well. First, in 2003 Tomassini et al. [7] published an analytic and numeric treatment of the problem, considering parameters typical for LWFA. The laser intensity is tuned such that the generated wake field is just below the threshold of wave breaking and self injection. This ensures that maximum charge is trapped at the density transition but the regular wake structure is preserved and guaranties a clean and reproducible electron beam. The parameters in their 2.5D PIC simulations are as follows: a 20 μm long density plateau at $2.1 \times 10^{19}\,\mathrm{cm}^{-3}$ is connected by a sharp density transition to a second plateau at $1.1 \times 10^{19}\,\mathrm{cm}^{-3}$ which is 150 μm long. The density transition occurs as a linear density drop over a length of 2 μm. The laser pulse that drives the wake field has an intensity of $2.5 \times 10^{18}\,\mathrm{W/cm}^2$ corresponding to a normalized amplitude a_0 of 1.3, a pulse duration of 17 fs (FWHM) and a spot size of 12 μm. This corresponds to a laser pulse energy of 110 mJ. With these parameters, a charge of 15 pC is injected and accelerated to an energy of 10 MeV with an energy spread of 5% and a normalized RMS emittance of 0.1-mm mrad. The simulation shows at this stage no sign of dephasing or laser pulse depletion, so further acceleration should be possible. The Rayleigh-range with the given spot size is approximately 400 μm. So even in the absence of self-guiding much higher energies should be attainable. The aspect of self-focusing is unfortunately also the weak spot of the study by Tomassini et al. because it is a well-known fact that self-focusing depends on the dimensionality of the system [8], and so does the process of transversal injection. Therefore, to be conclusive, a full 3D study is necessary in order to ensure that self-focusing does not lead to parasitic wave-breaking.

Still, with this point to be clarified, the results by Tomassini et al. are especially relevant within the context of the present work because LWS-20, the upgrade of LWS-10, with a planned on-target energy of up to 70 mJ will be able to reach the above-mentioned optimal intensity at just the same spot size of 12 μm due to the short pulse duration of 8 fs. Therefore, under the assumption that pump depletion is still negligible for the considered propagation distances, the results from [7] should be valid here as well.

Up to the present day, several publications treating this topic appeared showing certain refinements to the injection scheme. It was shown [9] that by replacing the second (low) density plateau by a slowly rising density ramp, dephasing can be omitted and the electron bunch always stays in the point of maximum accelerating gradient in the wake field, leading to larger final energy of the accelerated electron bunch. This density rise might also help to alleviate problems caused by beam-loading. However, several works point towards the possibility that gradually *lowering* the density in the low density region should lead to reduced energy spread [9–11]. Another recent study analyzes via 1D-PIC simulation the influence of amplitude and scale-length of the density transition on the amount of trapped charge with the basic conclusion that—at least in 1D geometry—more charge is trapped for larger amplitude and shorter scale-length of the transition. However, no information is given on how the trapped charge is phased in the acceleration field which, of course, is of utmost importance for the final electron beam parameters. To complete the list of theoretical and simulation results on the topic of electron injection into wake fields using sharp density transitions, the paper by Brantov et al. [12] must be taken into account where an analytical model based on Hamiltonian analysis is presented. And finally, as an outlook, the concept-paper by Petrillo et al. should be mentioned where the superior electron beam quality produced by this injection and acceleration scheme is used to realize an all-optical free electron laser. Although a preliminary result, it represents—if realizable—a very intriguing possibility for generating radiation with a wavelength on the 1 Å scale by replacing the usually employed undulator with a counter-propagating laser pulse.

Backed by all these encouraging theoretical results, proposals and attempts were made, as to how such a density transition could be realized. These proposed methods are: partially blocking a flow of low density plasma in the case of PWFA [6], shock fronts produced by colliding plasmas that are generated via laser-ionization of thin foils [7], and using a second laser at right angles with the driving beam that forms a parabolic density profile with steep edges by ionization and heating of an expanding plasma channel. Most likely, due to the involved experimental difficulties, of all these methods only the latter one was experimentally realized [13–16] but none of them yielded mono-energetic electron spectra. This can be attributed again to experimental difficulties with the rather complex setups, to non-optimal available laser parameters and to the rather larger deviations of the density-distributions produced in these experiments as compared to the ideal case considered in the theoretical studies.

So it seems that this scheme stands and falls with a reliable and straightforward way to produce step-like density profiles. In Sect. 3.2.6 such a way was presented for the first time to the knowledge of the author: shock fronts in supersonic gas flows. Shock fronts in supersonic gas jets are easily produced just by placing an obstacle like a knife edge into the gas jet. The shock front then naturally produces exactly the desired gas density profile: A high density region that is connected to a low density region by a short scale-length density jump as can be seen in Fig. 3.25. The scale-length is here on the order of the molecular mean free path which depends only on gas type (through the scattering cross on) and the number density. For a Helium jet of typical conditions, it is approximately 500 nm—much shorter than the plasma wavelength. Since the non-relativistic plasma wave length is given by $\lambda_P = 2\pi c\sqrt{(\varepsilon_0 m_e)/n_e e^2}$, and the molecular mean free path by $\lambda_{MFP} = 1/(\sqrt{2}\sigma n_G)$, the ratio between the two is given by

$$\frac{\lambda_P}{\lambda_{MFP}} = \frac{2\sqrt{2}\pi c\sigma}{e}\sqrt{\varepsilon_0 m_e}\sqrt{\frac{n_G}{\imath}} \tag{8.1}$$

Here, n_G is the neutral gas particle density, n_e is the electron density, σ is the molecular (or atomic) Van der Waals radius which amounts to 280 pm in the case of Helium and $\imath$ is the average degree of ionization of the atoms in the plasma. The rest of the symbols have their usual meaning. Figure 8.1 shows this relation for plasma densities between 10^{18} and 10^{20} cm^{-3} for Helium. Obviously, high plasma densities are beneficial in this scheme because they lead to transition scale-length much shorter than the plasma wave length. At a density of 2×10^{19} cm^{-3} the ratio between the two amounts to almost an order of magnitude.

Experimentally it has been shown in Sect. 4.4 that these shock fronts can easily be produced and visualized interferometrically. The shock fronts in the experiment are in a steady state and are reliably produced at the same location. Since the amplitude and position of the shock front only depends on the Mach number of the gas jet, the shock front is generated at the exact same location in a wide pressure range. The exact experimental characterization of these shock fronts as well as their application to laser driven electron acceleration are subject of ongoing and future work.

To summarize, it can be said that the concept of electron injection in non-breaking plasma waves by short-scale density transitions is a theoretically and numerically well studied and well-understood concept that shows several clear advantages as compared with electron injection via wave breaking. First and above

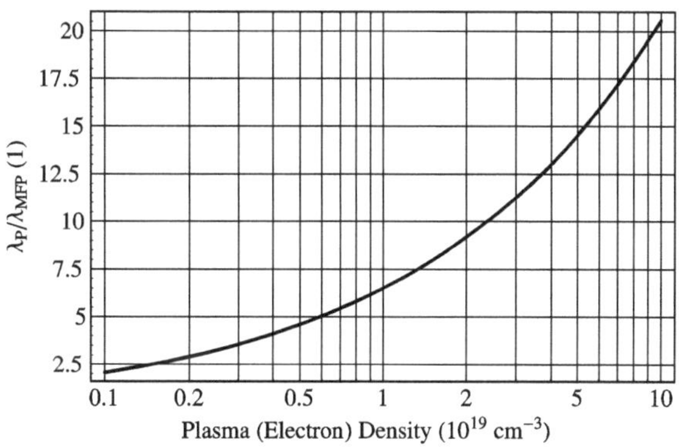

Fig. 8.1 Ratio between plasma wave period (assuming full ionization) and mean free path for Helium versus plasma (electron) density. High electron density lead to large ratio between mean free path (transition scale length) and plasma wave length

all, the injection of electrons happens at one well-defined spot which is expected to improve the shot-to-shot reproducibility of the electron beam by orders of magnitude. This claim is backed by encouraging experimental results obtained with triggered injection by counter-propagating laser pulses [1, 2]. By making important parameters as the amount of injected charge, the acceleration length and the process of dephasing experimentally easily accessible and tuneable, this new scheme can be expected to allow a leap in the development of user-friendly laser driven electron accelerators which are up to now still an object of basic research. For the first time, a straightforward, easy way to produce the density profiles called for by this method has been developed within this work. In the near future, this will allow the experimental investigation of electron injection in plasma waves below the wave breaking threshold, triggered by short-scale density transitions.

The general outlook to laser wakefield electron acceleration (LWFA) seems bright. Many experiments have proven that the exceedingly high acceleration gradients predicted in [17] exist and that they can be exploited for electron acceleration [18–40]. Maximum electron energies of up to 1 GeV have been demonstrated [19] with laser parameters that are surpassed by laser systems which are nowadays commercially available. Therefore it seems to be only a question of a few years before finally also the 10 GeV barrier will be broken.

Not only in terms of electron energy, also in terms of stability and reproducibility great improvements have been made [1–40]. Especially schemes for externally triggering self-injection like the one employing counter-propagating laser pulses [1, 2] or by density transitions offer the unique possibility of combining the advantages of self-injection with those of robust, close-to-one dimensional laser wakes. Of course, there remains much to be done. Further improvement of the electron beam stability together with larger repetition rate of the accelerator will allow more sophisticated experiments with the electron beam and the accumulation of larger statistics. Increasing the accelerated charge and especially the charge density due to ultrashort electron pulse durations will allow new ways of producing short wavelength radiation. The ultrashort duration might also allow time resolved diffraction studies of unprecedented resolution. And finally, the staging of multiple wake-field accelerators remains to be shown. Only by staging, the electron energy range relevant for fundamental particle physics will become accessible.

Considering all this, truly application-ready laser-driven electron accelerators seem to be within reach offering the prospect of table-top-scale experiments with electron beams in the GeV-range. However, there is still a lot of interesting work to be done in laser driven electron acceleration and relativistic laser-plasma physics so that these areas will stay attractive and will keep flourishing for years to come.

References

1. Faure, J., Rechatin, C., Norlin, A., Lifschitz, A., Glinec, Y., Malka, V.: Controlled injection and acceleration of electrons in plasma wakefields by colliding laser pulses. Nature **444**, 737 (2006)

2. Rechatin, C., Faure, J., Ben-Ismail, A., Lim, J., Fitour, R., Specka, A., Videau, H., Tafzi, A., Burgy, F., Malka, V.: Controlling the phase–space volume of injected electrons in a laser–plasma accelerator. Phys. Rev. Lett. **102**(16), 164801 (2009)
3. Bulanov, S., Naumova, N., Pegoraro, F., Sakai, J.: Injection into the wave acceleration phase due to nonlinear wake wave breaking. Phys. Rev. E. **58**(5), R5257 (1998)
4. Suk, H., Barov, N., Rosenzweig, J.B., Esarey, E.: Plasma electron trapping and acceleration in a plasma wake field using a density transition. Phys. Rev. Lett. **86**(6), 1011–1014 (2001)
5. Suk, H.: Electron acceleration based on self-trapping by plasma wake fields. J. Appl. Phys. **91**(1), 487–491 (2002)
6. Thompson, M.C., Rosenzweig, J.B., Suk, H.: Plasma density transition trapping as a possible high-brightness electron beam source. Phys. Rev. ST Accel. Beams **7**(1), 011301 (2004)
7. Tomassini, P., Galimberti, M., Giulietti, A., Giulietti, D., Gizzi, L., A., Labate, L., Pegoraro, F.: Production of high-quality electron beams in numerical experiments of laser wakefield acceleration with longitudinal wave breaking. Phys. Rev. ST Accel. Beams **6**(12), 121301 (2003)
8. Pukhov, A., Meyer-ter Vehn, J.: Relativistic magnetic self-channeling of light in near-critical plasma: Three-dimensional particle-in-cell simulation. Phys. Rev. Lett. **76**(21), 3975–3978 (1996)
9. Suk, H., Kim, C., Kim, G.H., Kim, J.U., Ko, I.S., Lee, H.J.: Energy enhancement in the self-injected laser wakefield acceleration using tapered plasma densities. Phys. Lett. A. **316**, 233 (2003)
10. Suk, H., Lee, H.J., Ko, I.S.: Generation of high-energy electrons by a femtosecond terawatt laser propagating through a sharp downward density transition. J. Opt. Soc. Am. B **21**(7), 1391 (2004)
11. Suk, H., Lee, H.J.: Tapering effects of a plasma density in laser wakefield accelerations. J. Kor. Phys. Soc. **44**(5), 1318 (2004)
12. Brantov, A.V., Esirkepov, T.Zh., Kando, M., Kotaki, H., Bychenkov, V.Yu., Bulanov, S.V.: Controlled electron injection into the wake wave using plasma density in homogeneity. Phys. Plasmas **15**(7), 073111 (2008)
13. Kim, J.U., Hafz, N., Suk, H.: Electron trapping and acceleration across a parabolic plasma density profile. Phys. Rev. E **69**(2), 026409 (2004)
14. Chien, T.-Y., Chang, C.-L., Lee, C.-H., Lin, J.-Y., Wang, J., Chen, S.-Y.: Spatially localized self-injection of electrons in a self-modulated laser-wakefield accelerator by using a laser-induced transient density ramp. Phys. Rev. Lett. **94**(11), 115003 (2005)
15. Jang, H., Kim, J., Yoo, S., Hur, M.S., Suk, H., Cho, M.-H., Namkung, W.: Plasma channel generation for electron acceleration with a laser-induced density gradient. J. Kor. Phys. Soc. **50**(5), 1466 (2007)
16. Kim, J., Jang, H., Yoo, S., Hur, M., Hwang, I., Lim, J., Kulagin, V., Suk, H., Choi, I. W., Hafz, N., Kim, H.T., Hong, K.-H., Yu, T.J., Sung, J.H., Jeong, T.M., Noh, Y.-C., Koh, D.- K., Lee, J.: Quasi-monoenergetic electron-beam generation using a laser accelerator for ultra-short x-ray sources. Jour. Kor. Phys. Soc. **51**(1), 397 (2007)
17. Tajima, T., Dawson, J.M.: Laser electron accelerator. Phys. Rev. Lett. **43**(4), 267 (1979)
18. Hamster, H., Sullivan, A., Gordon, S., White, W., Falcone, R.W.: Subpicosecond, electromagnetic pulses from intense laser–plasma interaction. Phys. Rev. Lett. **71**(17), 2725–2728 (1993)
19. Nakajima, K., Kawakubo, T., Nakanishi, H., Ogata, A., Kitagawa, Y., Kodama, R., Mima, K., Shiraga, H., Suzuki, K., Yamakawa, K., Zhang, T., Kato, Y., Fisher, D., Downer, M., Tajima, T., Sakawa, Y., Shoji, T., Yugami, N., Nishida, Y.: Proof-of- principle experiments of laser wakefield acceleration using a 1 ps 10 tw nd:glass laser. In: Proceedings of AIP Conference Advanced Accelerator Concepts 145–155 (1995)
20. Downer, M.C., Siders, C.W., Fisher, D.F., LeBlanc, S.P., Rau, B., Gaul, E., Tajima, T., Babine, A., Stepanov, A., Sergeev, A.: Laser wakefield photon accelerator: optical diagnostics for the laser wakefield accelerator based on longitudinal interferometry. Bull. Am. Phys. Soc. **40**, 1862 (1995)

21. Marquès, J.R., Geindre, J.P., Amiranoff, F., Audebert, P., Gauthier, J.C., Antonetti, A., Grillon, G.: Temporal and spatial measurements of the electron density perturbation produced in the wake of an ultrashort laser pulse. Phys. Rev. Lett. **76**(19), 3566–3569 (1996)
22. Gorbunov, L.M., Kirsanov, V.I.: Excitation of plasma waves by an electromagnetic wave packet. SOV. Phys. JETP **66**, 290–294 (1987)
23. Bulanov, S.V., Kirsanov, V.I., Sakharov, A.S.: Excitation of ultrarelativistic plasma waves by pulse of electromagnetic radiation. JETP Lett. **50**, 198–201 (1989)
24. Sprangle, P., Esarey, E., Ting, A.: Nonlinear theory of intense laser–plasma interactions. Phys. Rev. Lett. **64**(17), 2011–2014 (1990)
25. Sprangle, P., Esarey, E., Ting, A.: Nonlinear interaction of intense laser pulses in plasmas. Phys. Rev. A **41**, 4463–4469 (1990)
26. Coverdale, C.A., Darrow, C.B., Decker, C.D., Mori, W.B., Tzeng, K-C., Marsh, K.A., Clayton, C.E., Joshi, C.: Propagation of intense subpicosecond laser pulses through underdense plasmas. Phys. Rev. Lett. **74**(23), 4659–4662 (1995)
27. Nakajima, K., Fisher, D., Kawakubo, T., Nakanishi, H., Ogata, A., Kato, Y., Kitagawa, Y., Kodama, R., Mima, K., Shiraga, H., Suzuki, K., Yamakawa, K., Zhang, T., Sakawa, Y., Shoji, T., Nishida, Y., Yugami, N., Downer, M., Tajima, T.: Observation of ultrahigh gradient electron acceleration by a self-modulated intense short laser pulse. Phys. Rev. Lett. **74**(22), 4428–4431 (1995)
28. Modena, A., Najmudin, Z., Dangor, A.E., Clayton, C.E., Marsh, K.A., Joshi, C., Malka, V., Darrow, C.B., Danson, C., Neely, D., Walsh, F.N.: Electron acceleration from the breaking of relativistic plasma waves. Nature **377**(6550), 606–608 (1995)
29. Wagner, R., Chen, S.-Y., Maksimchuk, A., Umstadter, D.: Electron acceleration by a laser wakefield in a relativistically self-guided channel. Phys. Rev. Lett. **78**(16), 3125–3128 (1997)
30. Moore, C.I., Ting, A., Krushelnick, K., Esarey, E., Hubbard, R.F., Hafizi, B., Burris, H.R., Manka, C., Sprangle, P.: Electron trapping in self-modulated laser wakefields by raman backscatter. Phys. Rev. Lett. **79**(20), 3909–3912 (1997)
31. Ting, A., Moore, C.I., Krushelnick, K., Manka, C., Esarey, E., Sprangle, P., Hubbard, R., Burris, H.R., Fischer, R., Baine, M.: Plasma wakefield generation and electron acceleration in a self-modulated laser wakefield accelerator experiment. Phys. Plasmas **4**(5), 1889–1899 (1997)
32. Santala, M.I.K., Najmudin, Z., Clark, E.L., Tatarakis, M., Krushelnick, K., Dangor, A.E., Malka, V., Faure, J., Allott, R., Clarke R., J.: Observation of a hot highcurrent electron beam from a self-modulated laser wakefield accelerator. Phys. Rev. Lett. **86**(7), 1227–1230 (2001)
33. Malka, V., Fritzler, S., Lefebvre, E., Aleonard, M.-M., Burgy, F., Chambaret, J.-P., Chemin, J.-F., Krushelnick, K., Malka, G., Mangles, S.P.D., Najmudin, Z., Pittman, M., Rousseau, J.-P., Scheurer, J.-N., Walton, B., Dangor, A.E.: Electron acceleration by a wake field forced by an intense ultrashort laser pulse. Science **298**(5598), 1596–1600 (2002)
34. Hafz, N.A.M., Jeong, T.M., Choi, I.W., Lee, S.K., Pae, K.H., Kulagin, V.K., Sung, J.H., Yu, T.J., Hong, K.-H., Hosokai, T., Cary, J.R., Ko, D.-K., Lee, J.: Stable generation of gev-class electron beams from self-guided laser–plasma channels. Nat. Phot. **2**, 571 (2008)
35. Faure, J., Glinec, Y., Pukhov, A., Kiselev, S., Gordienko, S., Lefebvre, E., Rousseau, J.-P., Burgy, F., Malka, V.: A laser–plasma accelerator producing monoenergetic electron beams. Nature **431**, 541 (2004)
36. Geddes, C.G.R. et al.: High quality electron beams from a laser wakefield accelerator using plasma-channel guiding. Nature **431**, 538 (2004)
37. Mangles, S.P.D., Murphy, C.D., Najmudin, Z., Thomas, A.G.R., Collier, J.L., Dangor, A.E., Divall, E.J., Foster, P.S., Gallacher, J.G., Hooker, C.J., Jaroszynski, D.A., Langley, A.J., Mori, W.B., Norreys, P.A., Tsung, F.S., Viskup, R., Walton, B.R., Krushelnick, K.: Monoenergetic beams of relativistic electrons from intense laser plasma interactions. Nature **431**, 535 (2004)
38. Hidding, B., Amthor, K.-U., Liesfeld, B., Schwoerer, H., Karsch, S., Geissler, M., Veisz, L., Schmid, K., Gallacher, J.G., Jamison, S.P., Jaroszynski, D., Pretzler, G., Sauerbrey, R.:

Generation of quasimonoenergetic electron bunches with 80-fs laser pulses. Phys. Rev. Lett. **96**(10), 105004 (2006)

39. Leemans, W.P., Nagler, B., Gonsalves, A.J., Toth, Cs., Nakamura, K., Geddes, C.G.R., Esarey, E., Schroeder, C.B., Hooker, S.M.: Gev electron beams from a centimetrescale accelerator. Nature Phys. **2**, 696 (2006)
40. Osterhoff, J., Popp, A., Major, Zs., Marx, B., Rowlands-Rees, T.P., Fuchs, M., Geissler, M., Hörlein, R., Hidding, B., Becker, S., Peralta, E.A., Schramm, U., Grüner, F., Habs, D., Krausz, F., Hooker, S.M., Karsch, S.: Generation of stable, lowdivergence electron beams by laser-wakefield acceleration in a steady-state-flow gas cell. Phys. Rev. Lett. **101**(8), 085002 (2008)

Chapter 9
Conclusion

The main objective of this work was to experimentally investigate electron acceleration in the bubble regime as described in the theoretical work by Pukhov and Meyer-ter-Vehn [1]. As detailed in Chap. 5, this regime poses stringent limits on laser pulse duration, laser pulse energy, focal spot diameter, plasma density, and propagation length in the plasma: first, and most importantly, sufficient intensity (depending on laser parameters between 10^{18} and $10^{19}\,\mathrm{W/cm^2}$) must be reached with the laser focal spot diameter being approximately twice as large as the pulse length. Then, the plasma density must be matched to the pulse duration such that the plasma wavelength is larger than double the pulse duration and the propagation distance in the plasma should match the maximum acceleration length which is proportional to the Rayleigh-range of the laser beam.

By the time this work was initiated, conventional Ti:sapphire laser technology was unable to produce the necessary ultrashort pulse durations with sufficient energy. Amplification gain narrowing lead to minimum pulse durations of approximately 35–40 fs while the lacking availability of large aperture crystals and strong enough pump lasers limited the amplified pulse energy to 1–2 J. Therefore, in this work an alternative approach was followed by using non-collinear optical parametric chirped pulse amplification (NOPCPA) to generate ultra broadband few-cycle light pulses with multi-TW power. In this laser parameter range, electron acceleration in the bubble regime is possible but the maximum acceleration length is in the $100\,\mu\mathrm{m}$ range thereby causing the need for well defined gas jets in this size-range to be used as targets in the experiments. Ideally, such a gas jet has a flat-top density profile in order to ensure clean and predictable interaction dynamics and sharp jet edges to precisely define the maximum acceleration length and to avoid any post-acceleration interaction of the electron bunch with the plasma.

These demands call for microscopic de Laval nozzles to be used for generating these jets. In Part I of this work this topic is investigated for the first time in the context of laser-plasma experiments.

K. Schmid, *Laser Wakefield Electron Acceleration*, Springer Theses,
DOI: 10.1007/978-3-642-19950-9_9, © Springer-Verlag Berlin Heidelberg 2011

The basic analytic considerations in Chap. 2 and the detailed numerical simulations presented in Chap. 3 shed light on the peculiar properties of supersonic gas flows in microscopic de Laval nozzles. Thorough parameter studies allowed the optimized design of such nozzles as well as the prediction of the gas jet properties under experimental conditions. The most important results of these parameter studies are:

1. **De Laval nozzles generate collimated cold jets with sharp jet edges**. As the gas expands through the nozzle, the density decreases approximately one order of magnitude. In this way, most of the enthalpy of the gas is converted into kinetic energy within the nozzle, the confinement by the nozzle walls directs the gas flow into forward direction.
2. **The width of the density gradient at the jet-edge grows with nozzle-length and ratio between throat an exit diameter**. For sharp jet-edges, short nozzles and low expansion inside the nozzle are preferable.
3. **Density modulations decrease with nozzle length and angle**. The violent expansion of the gas close to the nozzle-throat causes density modulations that slowly decay as the gas propagates through the remaining part of the nozzle.
4. **Jet divergence decreases with increasing diameter ratio**. This is an expected result because the more enthalpy is already used up inside the nozzle the less there is to cause transversal acceleration of the gas once it has left the nozzle.
5. **The density at the nozzle exit scales according to a power-law with the backing pressure**. Approximately linear behavior is only given for high Knudsen numbers Kn (high pressure and/or large nozzles) and small pressure variations.
6. **Lowering the density or shrinking the nozzle leads to equivalent results**. In both cases, the gradient width and the jet divergence increase.
7. **Simulations reproduce well the measurement**. This holds true also in the case of small nozzles where wall-effects are strong.

This preparatory work allowed the construction of the smallest de Laval nozzles ever used in laser-plasma experiments. An experimental setup for gas jet characterization was built, some measurement results are presented in Chap. 4. The measurements confirm the simulation results and, hence, the functionality of microscopic de Laval nozzles.

With the gas target being ready for experiment, the second part of this work is dedicated to laser driven electron acceleration.

A short introduction to the analytic theory of laser-driven electron acceleration is given in Chap. 5 together with simulation results. The numeric simulations were conducted matching the experimental parameters and reproduce all the main properties of the bubble-regime as cavitation, electron-self injection, and acceleration and give useful information about optimum experimental conditions.

The main experimental novelty in this work besides the microscopic de Laval nozzles, is the unique light source Light Wave Synthesizer 10 (LWS-10). Development and improvement of LWS-10 up to the point where it could finally be used for experiments was an important part of this work and is briefly discussed in Chap. 6. The system is now up and running and available for experiments producing pulses with 8 fs duration (FWHM) and an energy of 50 mJ. The experimental setup that was constructed and used for electron acceleration is described in the same chapter.

The most important experimental findings, which are presented in detail in Chap. 7, are as follows:

1. **Mono-energetic electron bunches can be generated by few-cycle multi-TW laser pulses**. The ultrashort laser pulse duration of 8 fs allowed to self-inject and accelerate electrons in a laser driven plasma wave using a laser pulse energy of only 40 mJ.
2. **Low energy spread and divergence, low Thermal Background**. Clean mono-energetic electron bunches are produced comprising an almost background-free energy spectrum with a narrow mono-energetic peak having an energy spread down to 3% and a divergence as small as 5 mrad.
3. **Electron acceleration routinely operated at 10 Hz repetition rate**. Due to the low thermal background of the electron spectra and due to the low gas load on the vacuum system caused by the microscopic de Laval nozzles, the electron acceleration can be run at 10 Hz.
4. **Stable low energy electron beam**. By using supersonic gas jets with a diameter of only 150 μm, thereby restricting the interaction and acceleration length, a comparatively stable electron beam could be generated. Over hundreds of shots, the electron energy remained stable at 7.6 ± 1.6 MeV (RMS). Since the 10 Hz repetition rate allows to rather quickly build up meaningful statistics, this in principle would allow first experiments with the electron beam as, for example, time-resolved electron diffraction.
5. **Maximum electron energies up to 50 MeV** Maximum electron energies up to 50 MeV could be generated, but no stable operation could be achieved in this case, with shot-to-shot energy fluctuations being as large as 100%.

From these results it becomes clear that the most pressing issue to be solved is the lacking shot-to-shot reproducibility of the (high energy) electron beam. As a conclusion to this work, a promising route towards a more stable electron beam is proposed in Chap. 8. The proposed method exploits shock fronts in supersonic gas jets introduced in Sect. 4.4 to trigger the injection of electrons into the wake field thereby circumventing the instabilities introduced by self-injection through transversal wave breaking.

Reference

1. Pukhov, A., Meyer-Ter-Vehn, J.: Laser wake field acceleration: the highly non-linear broken-wave regime. Appl. Phys. B **74**:355 (2002)

Appendix A: Numeric Setup of the Fluid Flow Simulations

The major issue in the construction of microscopic nozzles lies in the fact that viscosity and boundary layer effects, which are of minor importance in the design of macroscopic supersonic gas jets, play an important role in the flow-formation. This implies that for correct prediction of the flow in this regime, the Navier–Stokes equations must be solved. Since, this is only possible numerically, the commercial code fluent was employed. The fluid was helium assumed to be a perfect gas. The problem was set up in 2D-axisymmetric with double precision accuracy. The mesh was generated with the mesh generator GAMBIT and was refined up to a point where the solution did not change any more with any further refinement. Several full 3D simulations were also carried out and it could be verified that the axisymmetric assumption is justified.

The boundary conditions were a pressure inlet at the high pressure side of the nozzle and a pressure outlet confining the volume outside the nozzle exit. The walls are adiabatic.

After the gas leaves the de Laval nozzle it emanates into a vacuum. Of course a vacuum cannot itself be simulated within a continuum model, so in order to avoid problems with dilute flows, the boundaries of the flow domain outside the nozzle have to be close enough to the nozzle exit to ensure high enough particle density everywhere. Since, the flow through the pressure outlet is supersonic over a major part of the outlet boundary, the actual boundary condition prescribed there does not influence the solution as long as the pressure prescribed at the exit boundary is small enough, see Sect. 3.2.5. In this case the flow does not depend on downstream values and the solver has to extrapolate upstream values from downstream values. To verify this, simulations with outlet pressures of 10, 1 and 0.1 Pa were made, yielding the expected result that within the main jet (95% of the mass flow), the difference in local density between those three cases is less than 1%. Only in the regions outside the main gas flow, where density and pressure are very low, considerable deviations occur. If not stated differently, in the following all simulations were conducted with a pressure of 1 Pa prescribed on the vacuum boundary. Similarly, effects of imperfect inlet boundary conditions (eg. constant

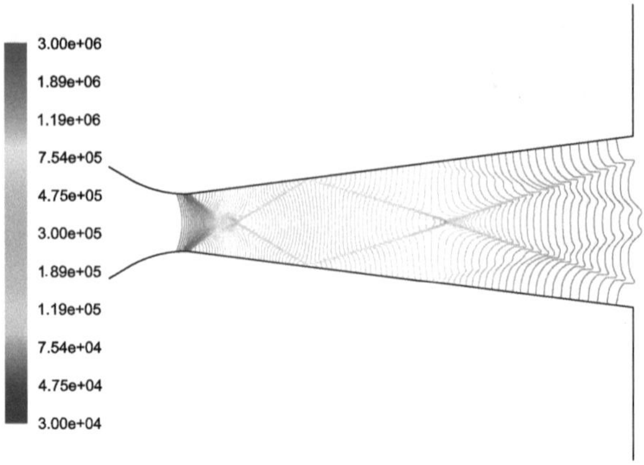

Fig. A.1 Pressure contours in a nozzle with a sharper exit-rounding at the throat. Medium: He, $d_C = 1\,\text{mm}$, $d_E = 3\,\text{mm}, L = 6\,\text{mm}$, $p_0 = 50$ bar. The rounded contour between the throat and the diverging section has a radius equal to half the throat diameter instead of four times it. This leads to the development of shock fronts that distort the flow quite strongly

flow parameters over the whole cross section) do not significantly influence the simulation results.

In order to achieve fast convergence, all problems were initialized in a two-step process. First the whole flow domain was initialized to inlet parameters, subsequently the pressure in the vacuum-domain was set to approximately 100 Pa. Then the f-m-g initialization function of fluent was used to obtain an initial state that shows already the main properties of the final solution, albeit with very poor accuracy. Sometimes convergence problems in the f-m-g-initialization made it necessary to start at higher background pressures of around 5,000 Pa and then stepwise lower the background pressure to the final value of 1 Pa.

A.0.1 The Optimal Mesh

In order to achieve numerical solutions which are independent of mesh-resolution and cell-geometry, a few quality factors must be observed during mesh-generation. The most obvious one is the resolution: The cell-size should be smaller (at least a factor of 2–3, better 5–10) than the smallest flow feature to be resolved. In the case of supersonic flows this can necessitate quite large mesh resolutions due to two reasons. First, in supersonic flows, shock fronts can occur which comprise very steep pressure gradients that have to be properly resolved. Such shock fronts can be seen in Fig. A.1.

The second reason is connected to the fact that we are interested here in the behavior of the subsonic, viscous boundary layer attached to the inner nozzle contour. Since, at the usual backing pressures in the range between 10 and 100 bar, the Reynolds numbers (calculated at the throat) are very high, it is clear that these boundary layers will be very small, on the order of $5\,\mu\text{m}$ in the throat.

The Reynolds number Re is defined as $Re = \rho w L/\mu$, with ρ being the density, w the velocity and μ the viscosity of the fluid. L is a characteristic length scale of the flow-geometry

To properly resolve these small scales, meshes have been used whose resolution normal to the walls is large enough to have at least 10 cells within the boundary layer. The main problem here is, that the boundary layer thickness is not a priori known, therefore it requires some experience to make a good initial guess. In principle, mesh adaptation can also be used for improving the boundary layer resolution during the simulation. However already in 2D the additional cells generated by one complete adaptation can be quite large and in 3D it is due to the large number of surface-cells usually impossible, or only at very high computational cost, to adapt large fractions of the boundary layer mesh. This problem is especially critical in the use of turbulence models like the k-ϵ, the k-ω, or the Reynolds Stress Model (RSM), since turbulent flows are significantly affected by the presence of walls. Obviously, the mean velocity field is affected through the no-slip condition that has to be satisfied at the wall. However, the turbulence is also changed by the presence of the wall in non-trivial ways. Very close to the wall, viscous damping reduces the tangential velocity fluctuations, while kinematic blocking reduces the normal fluctuations. Toward the outer part of the near-wall region, however, the turbulence is rapidly augmented by the production of turbulence kinetic energy due to the large gradients in mean velocity. The near-wall modelling significantly impacts the quality of numerical solutions, because the walls are the main source of vorticity and turbulence. Therefore, accurate representation of the flow in the near-wall region determines successful predictions of wall-bounded turbulent flows. The quantity Y^+ called the wall unit can be used to judge whether the boundary layer is sufficiently resolved for these turbulence models. It is defined as

$$Y^+ = \frac{\rho u_T y}{\mu} \tag{A.1}$$

where ρ is the local density, μ is the dynamic viscosity, y is the distance between the cell centroid and the wall for wall-adjacent cells and u_T is the tangential velocity. For the boundary layer to be properly resolved, Y^+ must be smaller than 1.

It can be seen in Fig. A.2 that for the given case ($d_C = 1\,\text{mm}$, $d_E = 3\,\text{mm}$, $L = 6\,\text{mm}$, $p_0 = 50\,\text{bar}$), Y^+ is smaller than one along the whole nozzle contour. It should be noted that due to the dependence of Y^+ on the density, it will vary roughly linear with the backing pressure. This means that the mesh corresponding to Fig. A.2a will be applicable in a range of approximately 10–100 bar.

The mesh is shown in Fig. A.2b with a closeup of the boundary layer mesh in the inset. Interestingly there is also a lower limit to Y^+, meaning that the boundary layer resolution can also be too high. This is probably due to the fact that the continuum approximation of fluid flows requires that the molecular mean free path is much smaller than the cell size. Therefore, it is interesting to note that for this reason it is necessary for low backing pressures or very small nozzles, to use

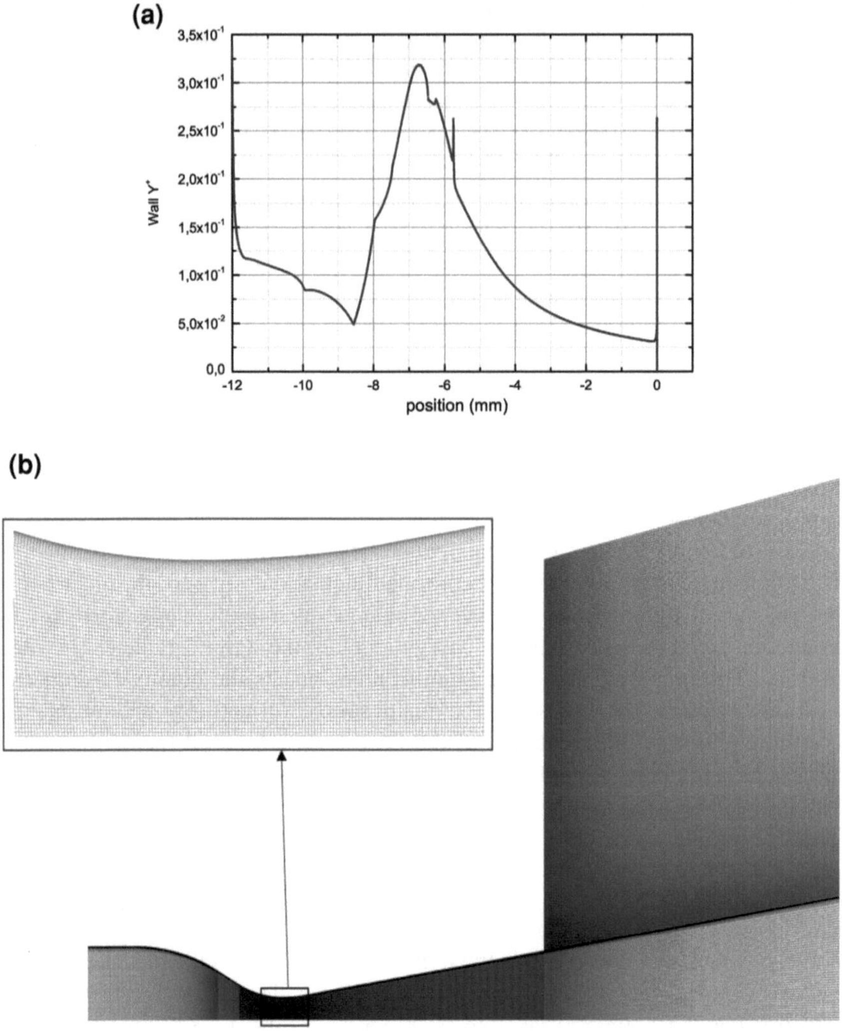

Fig. A.2 **a** Y^+ plotted along the walls of the nozzle at 50 bar. Y^+ is everywhere smaller than one, so the viscous boundary layer is properly resolved everywhere. **b** Typical mesh used in the simulations. The Inset shows the magnified nozzle throat. The increasing mesh resolution in radial direction towards the wall is clearly visible

meshes that actually contain less cells than for large nozzles or high pressures in order to obtain converging solutions. Of course the absolute resolution will still be higher for the smaller nozzles but on the smaller scale this can be accomplished with fewer cells. Mesh resolution is not the only quality factor that has to be satisfied by a mesh for the numeric solution of complicated nonlinear differential equations. As mentioned earlier, also the cell geometry plays an important role for the accuracy and stability of the numerical computation. Besides others, the most

Table A.1 Relationship between equi-angle skewness Q_{EAS} and mesh quality

Q_{EAS}	Quality
0	Equilateral (perfect)
$0 \leq Q_{QEAS} < 0.25$	Excellent
$0.25 \leq Q_{QEAS} < 0.5$	Good
$0.5 \leq Q_{QEAS} < 0.75$	Fair
$0.25 \leq Q_{QEAS} < 0.9$	Poor
$0.9 \leq Q_{QEAS} < 1$	Very poor (sliver)
1	Degenerate

important parameter that measures the geometric quality of a cell is the skewness. It is defined as the difference between the shape of the cell and the shape of an equilateral cell of equivalent volume. The equi-angle skewness, Q_{EAS} which was used here, is defined as

$$Q_{EAS} = \max\left\{\frac{\Theta_{max} - \Theta_{eq}}{180^\circ - \Theta_{eq}}, \frac{\Theta_{eq} - \Theta_{min}}{\Theta_{eq}}\right\} \quad (A.2)$$

Here Θ_{max} and Θ_{min} are the maximum and minimum angles (in degrees) between the edges of the element, and Θ_{eq} is the characteristic angle corresponding to an equilateral cell of similar form. For triangular and tetrahedral elements, $\Theta_{eq} = 60^\circ$. For quadrilateral and hexahedral elements, $\Theta_{eq} = 90^\circ$. By definition, $0 \leq Q_{QEAS} \leq 1$, where $Q_{EAS} = 0$ describes an equilateral element, and $Q_{EAS} = 1$ describes a completely degenerate element. The following table outlines the overall relationship between Q_{EAS} and element quality:

Highly skewed cells can decrease accuracy and destabilize the solution. For example, optimal quadrilateral meshes will have vertex angles close to 90°, while triangular meshes should preferably have angles of close to 60° and have all angles less than 90°. Especially for flows involving high pressure gradients—like the ones considered here—very high quality meshes are needed in order to obtain a converged solution. In the present case this limitation is so severe that it is practically impossible to use anything else than the map-scheme for meshing. This produces very low skewness, quadrilateral cell meshes with the drawback that there is only a limited possibility of cell size variation throughout the flow domain. This leads to larger numbers of cells as compared to meshes generated with more flexible schemes like the pave scheme in 2D for example. The displayed mesh in Fig. A.2b has for 95% of the cells a skewness <0.13, the worst cell has a skewness of 0.34.

A.0.2 Comparison of Numeric Flow Models

If one has to predict the flow properties of a given fluid with certain boundary conditions, the first question to answer is, which model to use. Although the Navier–Stokes (N–S) equations are complete and correct as long as the continuum approximation holds, it is usually impossible to run simulations on grid-resolutions high enough to resolve the smallest flow features. This is due to the fact that for turbulent flows these are represented by small scale eddies which are orders of magnitude smaller than the macroscopic flow features. However, they are still orders of magnitude larger than the molecular mean free path, and therefore represent a continuum effect which is correctly captured by the Navier–Stokes equations. To circumvent this problem several options are at disposition:

1. Ignore all viscosity and turbulence effects and solve the Euler equations. This is the Inviscid Model.
2. Ignore the small scales and solve the N–S equations on a coarse grid, capturing only macroscopic effects. This is the Laminar Model, because turbulence effects are ignored.
3. Rewrite all flow-parameters as the sum of a slowly varying average part and a rapidly (e.g. much faster than the macroscopic timescale) fluctuating turbulent part and introduce these parameters into the N–S equations. This allows to solve the N–S equations for the average parameters under the condition that there is a turbulence model at disposition that provides additional equations which allow to determine the additional parameters that have been introduced. In Fluent, several different turbulence models are implemented, which fall into three major categories: the k-ϵ models, the k-ω models and the RSM.
4. Resolve the larger turbulences (eddies) down to a given scale, and use a turbulence model for smaller scales. This is called Large eddy simulation (LES).
5. Direct numeric simulation (DNS), solving the N–S equations on a grid fine enough to capture even the smallest scales.

Ignoring all viscosity and turbulence effects, (Option 1) may seem a very crude approximation, but if the main interest lies for example, in the correct description of volume effects of supersonic and hypersonic flows, it can be a valid assumption. Option 2 can be taken if it is a priori clear that the effects of turbulence will be negligible for a given flow. This holds true for low Reynolds number laminar flows. At higher Reynolds numbers—the exact value depends on geometry—the microscopic turbulent fluctuations influence the macroscopic flow because they introduce additional transport effects for momentum and energy. This is taken into account in option 3, where so called two equation turbulence models, like the standard k-ϵ model and the k-ω model, assume that these turbulence driven transport effects are isotropic, and therefore use scalar equations to describe them. For stationary flows this mainly leads to an increased fluid viscosity, which can easily be understood considering that viscosity is responsible for transversal momentum transport and of course this transversal transport is enhanced by turbulent fluctuations of the fluid which happen as well in the transversal direction as in the longitudinal one, with respect to the macroscopic flow velocity. However, considering that these turbulence caused additional viscosity effects will play a significant role close to the confining walls of a flow where isotropy is always broken, it is clear that in order to capture the physics, a tensor–grade model has to be used, that allows for different transport terms in different directions. Such a model is the Reynolds Stress Model. Due to the additional equations that this model introduces as compared to the two equation models, it requires more memory space and usually also needs more iterations until convergence is reached. And finally, if macroscopic eddies develop in a flow, LES has to be used to correctly model the flow. Since, those large eddies are inherently time dependent, the problem has to be solved with a time dependent solver which adds largely to computational cost and post processing efforts. As mentioned above, option 4 is not available with current (and probably also future) computational resources.

Since the viscosity of the gas plays a decisive role in the formation of boundary layers, and therefore in the context of high mach number de Laval nozzles, the minimum requirement for describing the flow correctly is to solve the N–S equations without turbulence model. Since, we are also dealing with high Reynolds number flows, the influence of turbulence effects should also be checked for by means of one of the turbulence models. To underline this, it is noted that the Reynolds number calculated at the nozzle throat for a typical case ($d_C = 1\,\text{mm}$, $p_0 = 50$ bar) is 230,000. This compares to a critical Reynolds number of 2,300 for a cylindrical tube with smooth walls. At the critical Reynolds number, the flow changes from laminar to turbulent. Therefore, it is a priori clear that the flow in the de Laval nozzle is turbulent and the only remaining question is by how much the turbulence will influence the macroscopic flow. However, due to the large pressure gradients and the fact that the gas jet emanates into vacuum, it is sensible to assume that there will be no time dependent effects like the formation of large eddies within or outside of the nozzle. Therefore, no LES is necessary.

To address the question of whether or not turbulence plays a significant role, simulations with several different models have been performed. For the k-ϵ Model actually three different versions are implemented in Fluent, namely the Standard, (s-k-ϵ), the Realizable (r-k-ϵ) and the RNG (rng-k-ϵ) k-ϵ Models. For the k-ω Model there is one more version called the k-ω-SST Model, where SST stands for shear stress tensor. All of these models were run on a test case, which is a nozzle with the parameters $d_C = 1\,\text{mm}$, $d_E = 3\,\text{mm}, \text{L} = 8\,\text{mm}$ and $p_0 = 50$ bar.

For the Laminar model only the "Boundary Slip" option exists, which allows to implement the velocity slip and temperature jump that occurs at Knudsen numbers (see Sect. 2.1.2) larger than 0.003 or so, [1]. Since, the Knudsen number for 50 bar backing pressure is everywhere in the nozzle smaller than 10^{-5} (see Sect. 3.1.2), this option was not activated.

For the three k-ϵ Models the "Enhanced Wall Treatment" option was activated, which is the appropriate choice if the boundary layer is properly resolved ($Y^+ \sim 1$), as is the case here. (see Fig. A.2). Also the option "Pressure Gradient Effects" was activated because this improves the model for boundary layers affected by large pressure gradients.

In order to compare the different turbulence models, density plots for the laminar model (Fig. A.3), the three types of the k-ϵ models (Fig. A.3), and the two different k-ω Models, (Fig. A.3) are shown. The RSM was also tested but did not

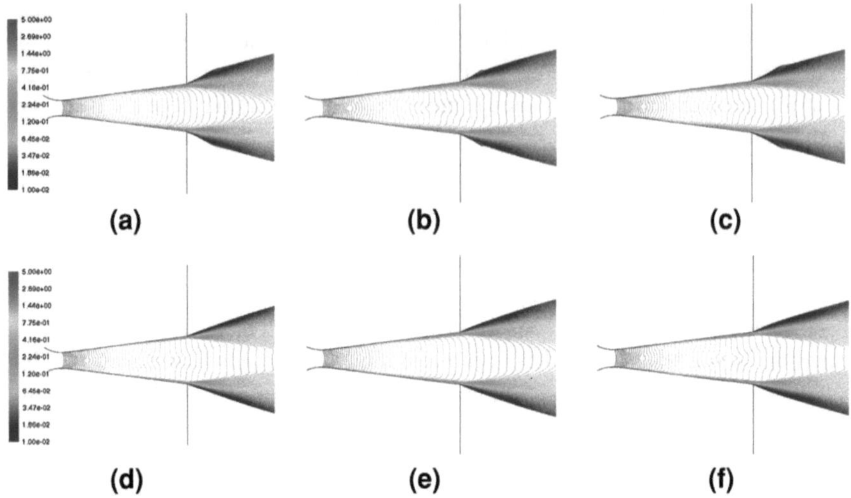

Fig. A.3 Comparison of several flow models with and without turbulence modelling. The plots show contours of constant density $(\mathrm{kg/m^3})$

converge, so the results are omitted here. All plots in Fig. A.3 have a logarithmic colormap.

It is obvious that the obtained solutions fall in two categories: those with and those without shock fronts. While the Standard-k-ϵ Model and the k-ω Model do not show any shock fronts, all the other models clearly show that a shock is generated close to the nozzle throat and propagates along the diverging section, and reflects off the wall.

Because of the fact that this study aims at the prediction of boundary layer thickness and density gradient width of (small) nozzles, also these boundary layer properties, as predicted by the three different models, were considered. The results for a nozzle with L = 8 mm, $d_C = 1$ mm and $d_E = 3$ mm are shown in Fig. A.4. The k-ω model has an additional option that improves the model for flows in the transitional regime for large Knudsen numbers. This option has been tested as well, results are given in the Fig. A.4. As can be seen, the predicted displacement thickness (see Sect. 3.1.3) and density gradient width differ considerably between the models. Figure A.4a shows the width of the density rise from the value at the wall to 90% of the maximum value of the respective transversal density profile. While the Laminar Model (blue line) marks the lower limit of all the calculations, the RSM gives the largest values. In between lie the different versions of the k-ϵ and the k-ω models. Interestingly, all k-ω models (red, green, yellow, cyan) agree quite well with each other even though they give different results regarding the shock fronts described above. While the k-ω (red line) and the k-ω-Transitional (yellow line) do not show any shock fronts (Fig. A.3e), the k-ω-SST (green line) and the k-ω-SST-Transitional (cyan line) do show them. (Fig. A.3f). However, this is different for the k-ϵ models. The two versions that show shock fronts,

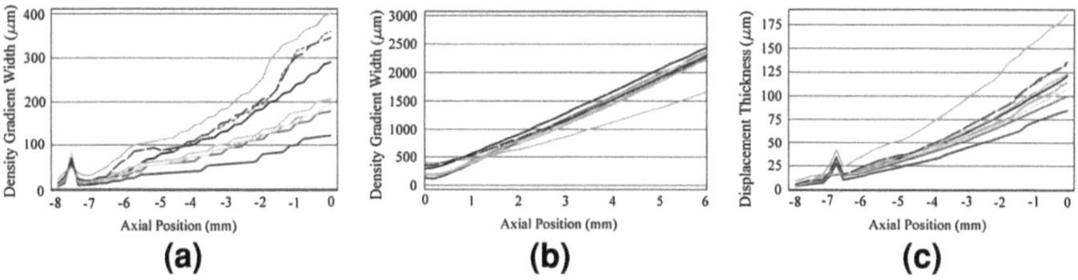

Fig. A.4 Comparison of turbulence models regarding boundary layer effects. *Black*: Standard-k-ϵ, short *dashes*: k-ϵ-Realizable, long *dashes*: k-ϵ-RNG, *red*: k-ω, *green*: k-ω-SST, *yellow*: k-ω-Transitional, *cyan*: k-ω-SST-Transitional, *blue*: Laminar, *grey*: Reynolds Stress Model

namely the k-ϵ–Realizable Model and the k-ϵ–RNG Model, give very similar results for the width of density gradient along the nozzle wall. The Standard-k-ϵ, that doesn't show shock fronts, also predicts a different density gradient width at the nozzle exit.

The gradient width of the free jet was defined as the radial distance between two points with 10 and 90% of the maximum density of a given radial density profile. Apart from the different start values caused by the differing values for the density gradient width at the exit of the nozzle, all the models except the RSM show very similar behavior, namely a linear increase of the gradient width starting approximately 0.3 mm after the nozzle exit. The deviating prediction of the RSM is most likely due to the above mentioned fact, that it did not converge in the free jet region of the flow.

It is interesting to note that within the first 0.3 mm after the nozzle exit, the density gradient stays fairly constant, making this region ideal for experiments that need a sharp, well defined and constant density gradient.

If one plots the displacement thickness inside the diverging part of the nozzle, as depicted in Fig. A.4c, a similar picture as for the density gradient width is obtained. Again, the Laminar Model gives the smallest and the RSM the largest values. The k-ϵ-Realizable and the k-ϵ-RNG model agree almost perfectly, the Standard-k-ϵ is a little bit off. The differences between the versions of the k-ω models are a bit more pronounced as for the density gradient width, however they are still fairly close to each other.

Regarding the possibility of rarefication effects within the k-ω Model it can be seen that there is a certain difference between the models with and without the transitional flow option, however the difference between the two versions of the k-ω models (k-ω and k-ω-SST) is larger than that.

To summarize it can be said that the tested flow models give results that fall in two categories, namely those with and those without (weak) shock fronts. Which of them corresponds closest to reality must be judged by experiment. Regarding the prediction of density gradient width and displacement thickness the models deviate quantitatively from each other, however all of them show similar trends. In both cases, the deviations between the models are too small to be detected by the interferometric measurements that were conducted in this study. Therefore the k-ω-SST model was used for the simulations because it is a well studied and popular turbulence model that has been verified on a large number of test cases and also shows favorable convergence behavior in the simulation.

Appendix B: Nozzle Designs

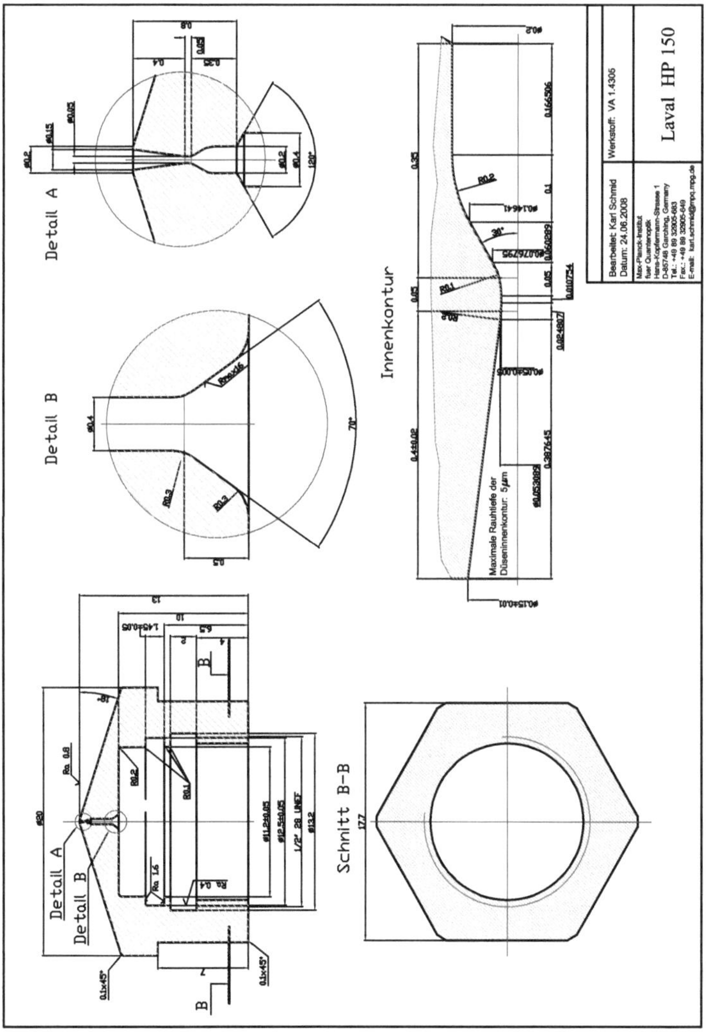

Fig. B.1 De Laval nozzle with exit diameter 150 μm

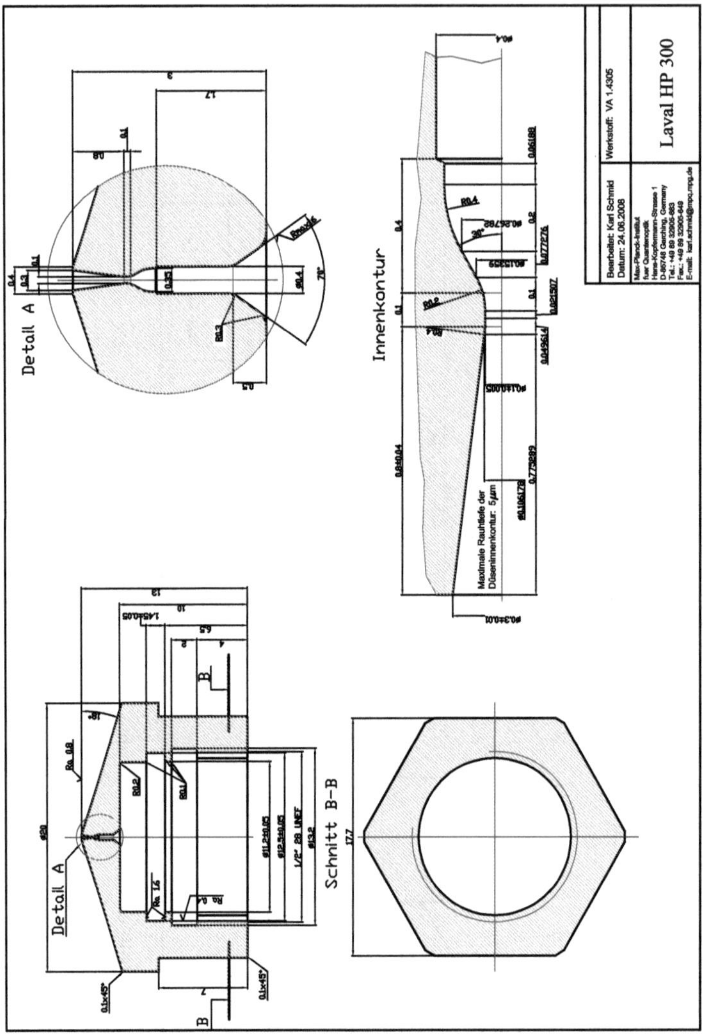

Fig. B.2 De Laval nozzle with exit diameter 300 μm

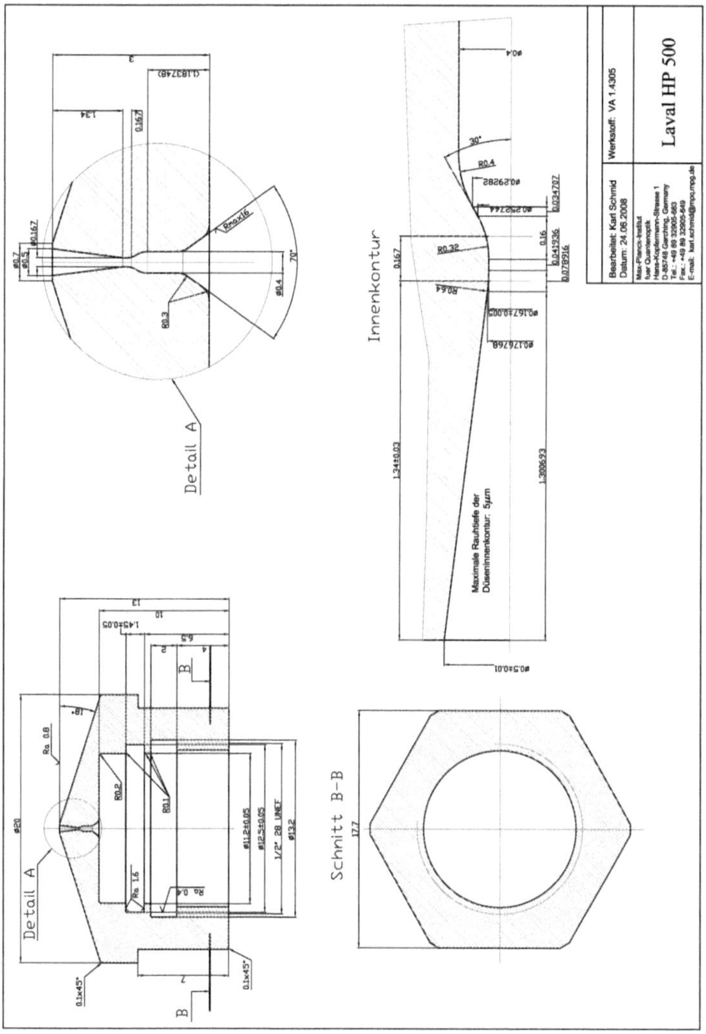

Fig. B.3 De Laval nozzle with exit diameter 500 μm

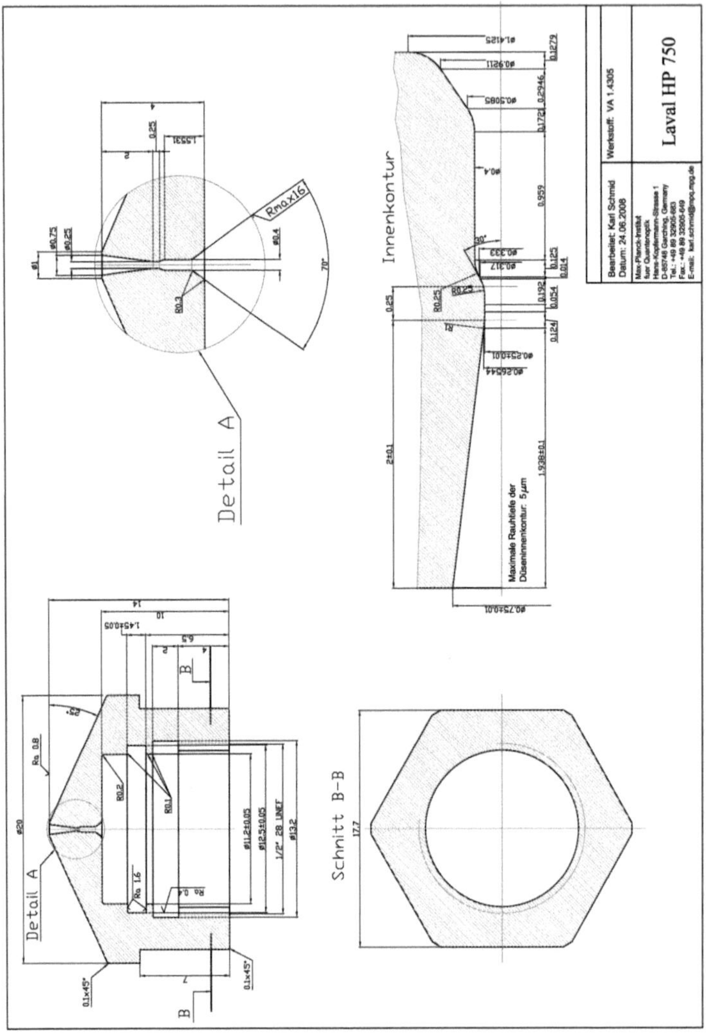

Fig. B.4 De Laval nozzle with exit diameter 750 μm

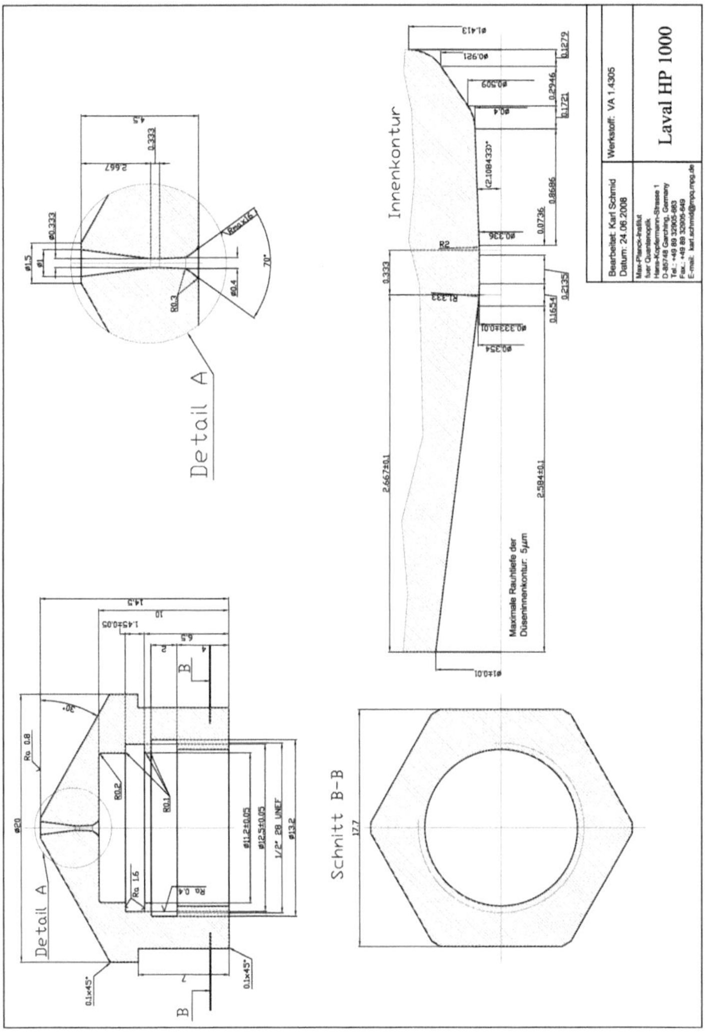

Fig. B.5 De Laval nozzle with exit diameter 1 mm

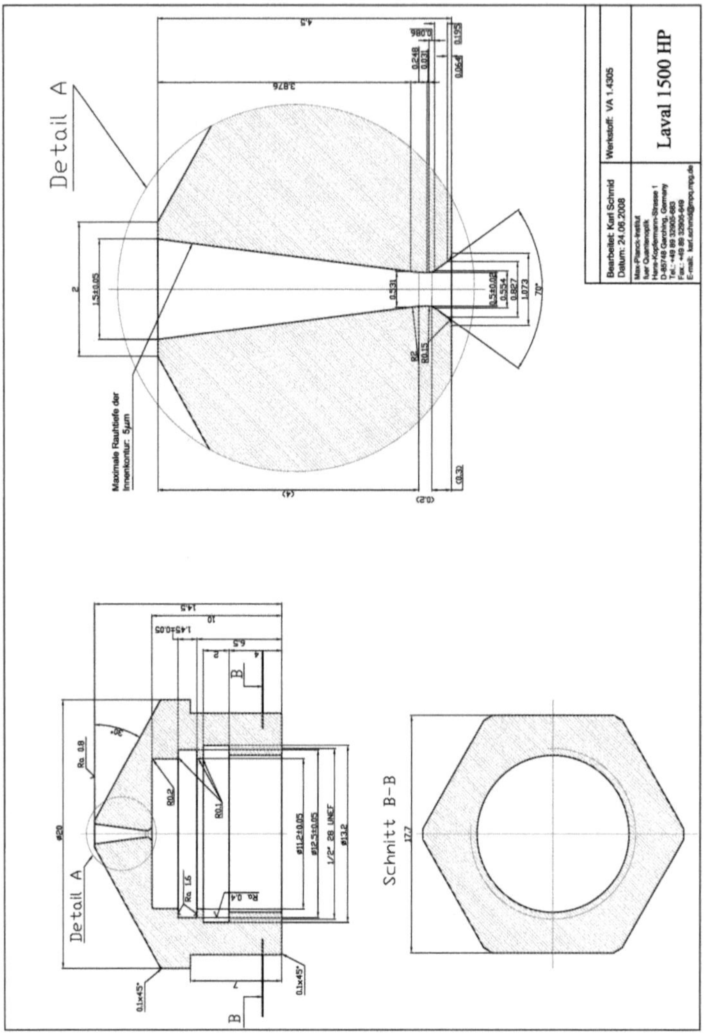

Fig. B.6 De Laval nozzle with exit diameter 1.5 mm

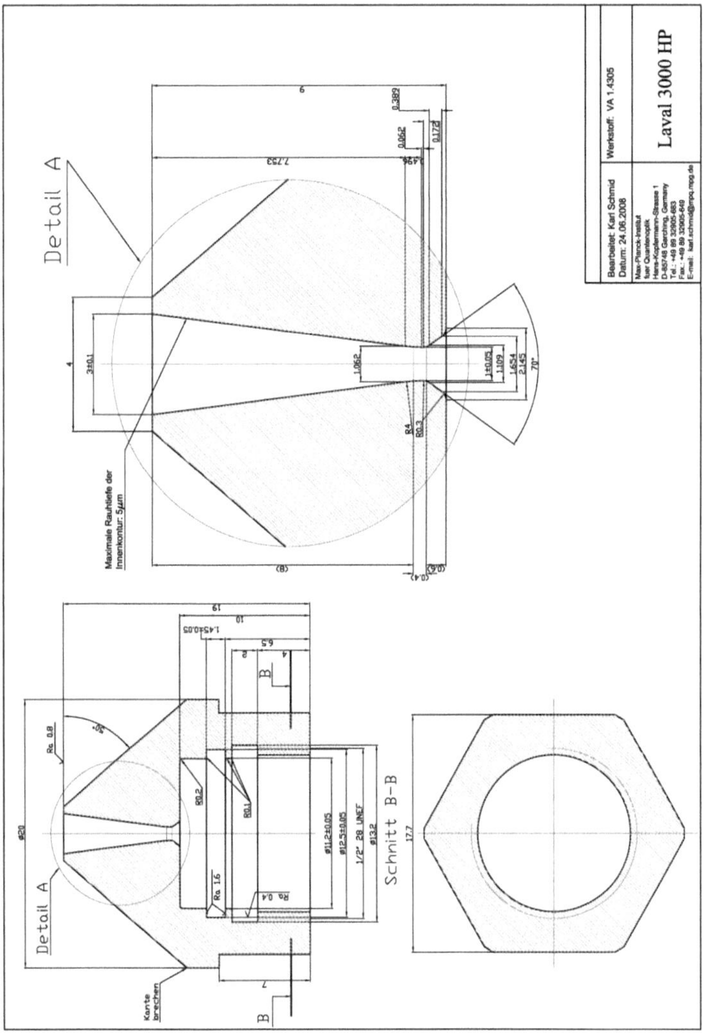

Fig. B.7 De Laval nozzle with exit diameter 3 mm

Reference

1. Borodulin, V.I., Rogalev, R.N., Slabospitsky, S.R. CORE: Compendium of Relations, arXiv:hep-ph/9507456

Curriculum Vitae

Contact

Gartenstrasse 24
D-71634 Ludwigsburg
Germany
E-mail: karl.schmid@gmx.eu

Personal data

Date of birth: April 27, 1978
Place of birth: Vienna, Austria
Citizenship: Austria

Professional memberships

Optical Society of America (OSA), German Physical Society (DPG), American Institute of Aeronautics and Astronautics (AIAA).

Activities

Founding member of and honorary work in the Southern Germany Student Chapter of the Optical Society of America (OSASG). Founding member of the international OSA network of students (IONS). Honorary work at the Austrian Boy Scouts and Girl Guides.

Education

Since 10/2004	PhD-student at the Max-Planck-Institute of Quantum Optics, Garching, Germany. Main research field: supersonic micro-flows and few-cycle laser-driven electron acceleration. Supervisor: Prof. Dr. Ferenc Krausz
10/03–10/04	Working on the diploma thesis "A Third Order Autocorrelator for the Characterisation of Ultrashort Laser Pulses" at the Vienna University of Technology. Supervisor: Prof. Dr. Ferenc Krausz
10/02–06/03	International Student at the Universidad Autónoma de Madrid, Spain
10/98–10/02	Study of physics at the Vienna University of Technology, Austria
10/97–05/98	Military service
09/92–06/97	TGM—Institute of Technology (engineering high school), Vienna, Austria Vehicle/Lightweight Construction; Graduated with distinction

Internships

09/00	Siemens AG, Vienna, Austria
07/99	J. Christof GmbH, Greater Vienna, Austria
06/98	J. Christof GmbH, Greater Vienna, Austria
07/97	J. Christof GmbH, Greater Vienna, Austria
07/96	J. Christof GmbH, Greater Vienna, Austria
07/95	OMV AG, Greater Vienna, Austria
07/94	OMV AG, Greater Vienna, Austria

MIX
Papier aus verantwortungsvollen Quellen
Paper from responsible sources
FSC® C105338

If you have any concerns about our products,
you can contact us on
ProductSafety@springernature.com

In case Publisher is established outside the EU,
the EU authorized representative is:
Springer Nature Customer Service Center GmbH
Europaplatz 3, 69115 Heidelberg, Germany

Printed by Libri Plureos GmbH
in Hamburg, Germany